Academic Practice - a Handbook for Physicists and Engineers Involved in Biomedical Research and Teaching

Editors

Peter R. Hoskins, Edinburgh
Stephen F. Keevil, London

Authors

Richard A. Black, Glasgow
Sally Bradley, York
Alicia El Haj, Stoke-on-Trent
Patrick A. Finlay, London
Jamie Harle, London
Peter R. Hoskins, Edinburgh
Stephen F. Keevil, London
Timm Krüger, Edinburgh
Catherine Lillie, Hull
Keith M. McCormack, Sheffield
Alison Robinson-Canham, York
Michael A. Smith, York and Wigtown
Wendy B. Tindale, Sheffield

Contributors

Richard A. Black
Senior Lecturer in Bioengineering
Department of Biomedical Engineering
University of Strathclyde
Glasgow, UK.
Editor-in-Chief, Medical Engineering & Physics.
Email: richard.black@strath.ac.uk

Sally Bradley
Academic Lead: Professional Learning and Development
The Higher Education Academy
York, UK.
Email: Sally.Bradley@heacademy.ac.uk

Alicia El Haj
Professor of Cell Engineering
Institute for Science and Technology in Medicine
Keele University
Stoke-on-Trent, UK.
Email: a.j.el.haj@keele.ac.uk

Patrick A. Finlay
Chief Executive, Institute of Measurement and Control
87 Gower Street London, UK.
Founder and managing director MediMaton Ltd.
Email: patrick.finlay@instmc.org

Jamie Harle
Senior Teaching Fellow
Medical Physics and Biomedical Engineering
Faculty of Engineering Science
University College London
London, UK.
Email: j.harle@ucl.ac.uk

Peter R. Hoskins
Professor of Medical Physics and Biomechanics
Centre for Cardiovascular Science
& Bioengineering Institute
University of Edinburgh
Edinburgh, UK.
Email: p.hoskins@ed.ac.uk

Stephen F. Keevil
Head of Medical Physics
Guy's and St Thomas' NHS Foundation Trust
London UK.
Professor of Medical Physics
King's College London
London, UK.
Email: stephen.keevil@kcl.ac.uk

Timm Krüger
Chancellor's Fellow
School of Engineering
University of Edinburgh
Edinburgh, UK.
Email: timm.krueger@ed.ac.uk

Catherine Lillie
Academic Practice Advisor
University of Hull, Hull, UK.
Email: c.lillie@hull.ac.uk

Keith M. McCormack
Business Development Manager
Insigneo Institute for *in silico* Medicine
University of Sheffield
Pam Liversidge Building, Mappin St
Sheffield, UK.
Email: k.m.mccormack@sheffield.ac.uk

Alison Robinson-Canham
Assistant Director - Professional Practice
The Higher Education Academy
York, UK.
Email: alison.robinson-canham@heacademy.ac.uk

Michael A. Smith
Emeritus Professor of Medical Science
Managing Partner - Harper Keeley LLP
International Associate - Leadership Foundation for
Higher Education
York and Wigtown, UK.
Email: ProfMASmith@gmail.com

Wendy B. Tindale
Scientific Director of Medical Imaging & Medical
Physics, Sheffield Teaching Hospitals
Clinical Director of NIHR Devices for Dignity
Healthcare Technology Co-operative
Professor of Medical Physics, Insigneo Institute
for *in silico* Medicine
Sheffield, UK.
Email: Wendy.Tindale@sth.nhs.uk

© Institute of Physics and Engineering in Medicine 2017
Fairmount House, 230 Tadcaster Road
York YO24 1ES
ISBN 978-1-903613-63-4

Published by the Institute of Physics and Engineering in Medicine
Fairmount House, 230 Tadcaster Road, York YO24 1ES

Prepared by:
The Charlesworth Group, Huddersfield, UK. www.charlesworth-group.com

Contents

Author biographies

Richard Black (BSc, PhD, PGCert, CSci, CEng, FIMechE, FIPEM, FHEA) is a biomedical engineer with over 25 years' experience in the field. He graduated BSc (Hons) Engineering Science from the University of Edinburgh, and gained his PhD from the Faculty of Medicine at the University of Liverpool, where he was lecturer in Medical Device Design. He is currently academic director of the UK Engineering & Physical Sciences Research Council (EPSRC) Centre for Doctoral Training in Medical Devices and Health Technologies at the University of Strathclyde. A Chartered Scientist and Engineer, Dr Black is Fellow of the Institution of Mechanical Engineers and the Institute of Physics and Engineering in Medicine, and Editor-in-Chief of *Medical Engineering & Physics*, one of three journal publications of the Institute (www.medengphys.com).

Sally Ann Bradley (PhD, MSc, BSc (Hons), PFHEA, SFSEDA) studied Computing and Systems with the Open University graduating in 1998 and Strategic Information Systems Management at Sheffield Hallam University, where she is currently an Honorary Professor in the Faculty of Health and Wellbeing. She is the Academic Lead: Professional Learning and Development at the Higher Education Academy (HEA). Prior to joining the HEA, Sally was a Principal Lecturer at Sheffield Hallam University and as Professional Recognition Adviser, she developed and led the HEA accredited CPD Scheme. Previously, Sally has worked as an Educational Developer and Pedagogic Researcher, featuring in the 2008 Research Assessment Exercise (RAE). She sat on the Researcher Concordat and on the Research Excellence Framework (REF) reading panel.

Alicia El Haj (MSc, PhD, FRSB, FEAMBES) is Professor of Cell and Tissue Engineering at Keele University. She is a leading figure in Regenerative Medicine and has been involved in bringing together interdisciplinary groups within biomedicine, physical sciences and engineering interested in aspects of cell and tissue engineering. She has been a founder Director of The Institute of Science & Technology in Medicine (ISTM), one of the co-directors of the EPSRC Centre for Innovative Manufacturing Centre in Regenerative Medicine and a collaborator in the AR UK Tissue Engineering Centre. She was European President of TERMIS (2012-15), and is currently Chair of the Bioengineering Society. She has contributed widely to funding and assessment panels including the 2008 and 2014 REF panels. In 2012, she was awarded a Royal Society Merit Award, and in 2015 an MRC Suffrage Award for efforts to promote women in the sciences. She has published over 250 journal papers.

Patrick Finlay (BSc, PhD, CEng, FIMechE) is a serial entrepreneur and biomedical engineer. His speciality area is medical robotics and he was the founder of Europe's first medical robot company in 1993. Since then, he has founded and partnered in five other medical device companies, mostly involving image-guided robotics. He has an engineering degree from the University of Birmingham and a PhD in advanced automation from the University of Strathclyde. He is the author of some 20 patents and a substantial number of papers. From 2012 to 2015, he was Chairman of the Biomedical Engineering Association. He is currently Chief Executive of the Institute of Measurement and Control, a Fellow and Trustee Board member of the Institution of Mechanical Engineers, and a member of the Royal Academy of Engineering's forum for biomedical engineering.

Jamie Harle (BSc, MSc, DPhil, MSc, PGCert, CSci, MIPEM, MInstP, FHEA) graduated from Imperial College London in physics and then specialised in medical physics, completing a Masters and DPhil at the University of Oxford. After postdoctoral work at UCL, his medical physicist hospital training was at Cambridge. He has worked for the last 10 years as a teaching-specialist lecturer/teaching fellow in medical physics at The Open University, The University of Liverpool and University College London, delivering and developing a range of innovative educational programmes in physics applied to medicine. He is programme director of the MSc in Physics and Engineering in Medicine. Jamie was the recipient of the UCL Provost's Award for Leadership and Impact in Teaching in 2015, as well as winning the University of London Centre for Distance Education Teaching & Research Award in 2014.

Peter Hoskins (BA, MSc, PhD, DSc, PGCert, FIPEM, FInstP, FHEA) studied physics at Oxford University graduating in 1980. He has worked in Edinburgh since 1984, initially as a Hospital Physicist in the National Health Service where he became Consultant Physicist. He is currently Professor of Medical Physics and Biomechanics at Edinburgh University. Research interests have included development of ultrasound techniques for diagnosis of cardiovascular disease, patient specific modelling and elastography. He has published 135 refereed journal papers and is principal author of five books. He has contributed extensively to national (BMUS, IPEM) and international (IEC) guidance on ultrasound system performance testing and quality assurance.

Stephen Keevil (MA, MSc, PhD, ARCP, HonMRCR, CSci, FIPEM, CPhys, FInstP) studied physics at Oxford University and then trained as an NHS medical physicist. He completed a PhD on magnetic resonance spectroscopy and subsequently held a series of research and academic posts at King's College London, where he combined academic activity with support for the clinical MRI service. He was appointed Consultant Clinical Scientist and Head of Magnetic Resonance Physics at Guy's and St Thomas' NHS Foundation Trust in 2005 and made Professor of Medical Physics at King's in 2013. In 2017 he was promoted to Head of Medical Physics in the Trust. Professor Keevil has a portfolio of NHS and university responsibilities encompassing clinical service, research and education, and is Director of the King's Technology Evaluation Centre (KiTEC). Among other professional leadership roles, he is Past President of the Institute of Physics and Engineering in Medicine (IPEM) and has been a member of several NIHR funding panels.

Timm Krüger (Dipl-Phys, PhD, MIPEM) studied Physics at Bielefeld and Heidelberg University where he graduated in 2007. After finishing his PhD in Physics at the Max-Planck Institute in Dusseldorf in 2011, he worked as postdoc at Eindhoven University of Technology and University College London. Since 2013, he has been Chancellor's Fellow in the School of Engineering at the University of Edinburgh. His research interests include suspensions, interfacial phenomena, microfluidics and biophysical applications of blood flow. He has published 24 peer-reviewed journal papers and two physics textbooks.

Catherine Lillie (BA, MA, SFHEA) is an Academic Practice Advisor at the University of Hull. She has extensive experience of Higher Education at discipline, faculty and institutional level, working with all grades of staff on their personal, professional, career and teaching development. She is a Senior Fellow of the Higher Education Academy. Prior to joining the University of Hull, she worked in Researcher Development in the University of Manchester's Faculty of Science and Engineering, designing and running programmes for Research Staff and PGRs. She spent seven years as Executive Officer (Professional Development) at the AUA, where she ran the Postgraduate Certificate in Professional Practice (Higher Education Administration and Management), worked on the national CPD Framework project and designed, delivered and coordinated national initiatives and events.

Keith McCormack (BEng, PhD) graduated in 1975 from the University of Sheffield, and for 20 years worked in the electronics industry, designing telecommunications, broadcasting, computing, power and fluid control systems. In 1995, he returned to academia to design and conduct simulation and validation activities in *in silico* medical research. He eventually obtained his PhD in 2005, in 'Inhaled Therapy - Modelling and Practice' (well worth a read), and went on to run mostly EC-funded international research projects in cardiovascular medicine. In 2012, he joined the University of Sheffield's new Insigneo Institute for *in silico* Medicine, Europe's largest simulation-focused centre, where he now works in Project Management and Business Development, writing research proposals in a darkened room. His other interests are mostly French and consumable.

Alison Robinson-Canham (BA, MA, FCMI, FAUA) studied interdisciplinary Arts and Humanities with the Open University before pursuing postgraduate study in Post-Colonial Literature (University of Hull), Management (University of York) and Educational Research (Lancaster University). She has worked in higher education for over 20 years, making major contributions to many ground-breaking initiatives in professional development across the spectrum of academic and professional disciplines, in the UK and internationally. As Assistant Director for Professional Practice at the Higher Education Academy, Alison is responsible for Accreditation, Educational Excellence Awards and supporting the community of 85 000 Fellows.

Her own research and development interests are focused on the educational role of professional bodies, social theories of learning, and inclusivity in pedagogic practice. Alison is also an Independent Trustee of the Institute of Physics and Engineering in Medicine (IPEM).

Mike Smith (BSc, MSc, PhD, DSc, FIPEM, FBIR, FIKT, FInstP, FRSA) graduated in Physics from Bristol in 1973. He worked in medical physics in Edinburgh, London and Leeds where he was Professor and Director of Medical Physics, NHS R&D Director and Faculty Dean for Research in Medicine. The last 10 years of his full time career were spent as Deputy Vice Chancellor for Research and Enterprise at the University of Teesside then Pro Vice-Chancellor for Research and Knowledge Transfer at Sheffield Hallam University. He has published over 180 journal papers and was President of the BIR (1997-98). He has formed several medical device companies, and has been a non-executive director of several spin-out and start-up companies. He has served on numerous UK National Committees with responsibility for investment in new technologies. Now in pseudo-retirement, he has a range of activities in the charity, healthcare, academic and business fields. He is a managing partner at Harper Keeley LLP.

Wendy Tindale (OBE, MSc, PhD, CSci, FIPEM, ARCP) is Consultant Clinical Scientist and Scientific Director at Sheffield Teaching Hospitals. She holds a Professorship in Medical Physics at the University of Sheffield and nationally is Director of the NIHR 'Devices for Dignity' Healthcare Technology Co-operative. She has an interest in combining clinical academic excellence and patient benefit with business opportunities through the translation of innovative medical technologies and has worked with both public and private sectors to facilitate successful collaborations. Wendy has contributed to numerous national and international committees. She received a Beacon Award for outstanding contribution to healthcare and was honoured in the Health Service Journal's inaugural Inspirational Women awards for outstanding leadership in her field, becoming Healthcare Scientist of the Year in 2016.

Preface

This book is concerned with issues relevant to work in academia in Medical Physics, Medical Engineering and Bioengineering. It is aimed mainly at current and prospective university staff and PhD students, but there is also much that will be of interest to NHS staff, particularly those involved in teaching, working with academics or considering a clinical academic career, and similarly to staff in industry. There are four main sections: the academic career from PhD student to post-retirement, practical issues in research such as how to write a PhD thesis and project management, research funding in universities and also in the NHS and industry, and a final section on teaching. The perspective is UK-based and aimed at physicists and engineers working in medicine and healthcare. However, much of the general advice is relevant to science and engineering more widely and to an international audience. This is a practical 'how to' handbook; consequently, there are few references and the style is informal. Advice and guidance are often based on the personal experience of the authors and their colleagues, with the approach of 'what would I liked to have been told when I started in this area'. Therefore, it takes a realistic view of academic life, the pitfalls as well as the prizes.

Peter R. Hoskins, Stephen F. Keevil, Autumn 2016

Section 1

Academic careers

1 PhD student

Peter R. Hoskins and Stephen F. Keevil

1.1   Introduction

Much has been written on the subject of the PhD, much less on interdisciplinary PhDs and their particular issues. Anyone thinking of doing a PhD or already undertaking a PhD would do well to read the excellent text 'How to Get a PhD' by Philips and Pugh (Open University Press, now in its 6th edition). Any PhD is challenging and the student should be prepared to enjoy the experience of being challenged. The PhD is a rite of passage with its own initiation which is the PhD viva. PhD students remember their vivas for the rest of their lives. Academics will talk to each other about their own vivas: who had the shortest/longest viva (the shortest we have heard of was 45 minutes, the longest 7 hours), who had the viva where the supervisors answered the questions (supervisors are supposed to say nothing during the viva), who had the most aggressive examiner, who had the viva where the internal and external examiner spent the entire time arguing with each other. Submitting the PhD thesis, undertaking the viva, and attending the degree ceremony are all very significant events. The successful PhD student enters a new phase of life: they have developed enormously as a researcher and as a person; they have shown the ability to carry out independent research to a high level, and they are now part of an elite (around 1% of the UK population) that hold a doctoral degree and so can use the title 'Doctor' (the title 'Doctor' as used by members of the medical profession is usually a courtesy title, and hence medical doctors are not included in this figure). The successful PhD student should feel rightly proud of gaining their doctorate. Many students will have to overcome periods of self-doubt or even worse doubt from their supervisors. As well as academic ability and the ability to learn new technical skills, of equal or possibly greater importance is tenacity. In this respect the local environment and the support of family, friends and fellow students are important. This short section has briefly introduced the PhD. Below various issues relevant to the PhD process will be explored. Many of the issues are common across all PhDs, but some particular issues for physicists and engineers undertaking interdisciplinary PhDs are discussed.

1.2   The PhD degree

'PhD' stands for '*Doctor Philosophiae*' or 'Doctor of Philosophy'. The modern meaning of the PhD, as a degree involving original research and the production of a thesis, originated in Germany. It was adopted in the USA in the 1870s and in the UK in the early 1900s. The entry requirement for a PhD is usually a 2-1 or 1st in a relevant subject. The PhD degree usually takes 3 years of full time study.

Traditionally, PhDs were in single-field subjects such as physics or engineering. A PhD in interdisciplinary areas such as medical physics or bioengineering requires the student to undertake work in two or even three technical areas and to engage with the relevant biology or medicine. It is increasingly accepted that these PhDs take longer with 4 years being a more common time for completion, although funding bodies and universities have been slow to adapt to this model. Some supervisors unused to the challenges of interdisciplinary research have not helped by expecting students to hand in the thesis at the end of 3 years. It can be especially challenging if there is no funding beyond 3 years, although some 4-year studentships are now available. University deadlines for completion can be a challenge here too, with a 4-year absolute deadline now not uncommon apart from when there are extenuating circumstances. This is driven by the increasing use of PhD completion rates and times as university quality metrics by government agencies and independent commentators alike. Equally, it is common for NHS staff to undertake part-time PhDs, which may take 5 years or longer.

1.3   Student–supervisor relationship

The studentship belongs to the student, not the supervisors. The role of the supervisor is to start the student off in a direction which is at the cutting edge of research. Most PhDs have an initial direction sufficient to get the student started. The later stages of the research are usually not very well defined at the beginning. This is not bad planning on behalf of the supervisors; research by its nature is unpredictable and may not work at all, so it is harder to plan the later stages of a project. While the supervisors initiate the direction of the research, the student should have freedom to change direction later on. In other words, at some point, the PhD becomes owned by the student and the supervisors are there to provide guidance. This crossover from supervisors to student driving the project usually comes late in the first year or in the second year. There are significant variations in this: some PhDs are very prescriptive due to the constraints of the grant giving body or industrial funder. This is more like contract research and the question must be asked if this model of research is appropriate for a PhD, which should allow the student some freedom to follow their own research direction. Some students, especially those who have previous experience, take ownership of the PhD very early in the process, or even define their own PhD from initiation. A minority of students are unable to take ownership of the PhD, relying on their supervisors to direct throughout. Generally, this is not a good sign and foretells of problems in the exam. A common sign of this approach is apparent in the viva when the examiner asks 'why did you choose this particular methodology' to which the incorrect reply is 'because my supervisors told me'. Students with this approach are best advised early on that it is inappropriate. In the UK, students have an interim assessment at the end of the first year. This is a useful point to review progress and, if necessary, put corrective measures in place. Sometimes the panel judges that a student has made insufficient progress and that this is unlikely to change, in which case the studentship may be discontinued.

In this rare situation, there is usually the opportunity to write up the thesis for an MPhil (Master of Philosophy) degree.

It is also worth looking at issues of 'whose PhD is this' from the supervisor's point of view. Students will be supervised by an experienced researcher and will join that researcher's group. The supervisor provides funding, facilities and their own time for supervision and in return the student helps push forward the supervisor's research agenda. Sometimes students find themselves pressurised to support other aspects of this research agenda, beyond their own PhD project. For example, it may be that a contract researcher has left with a project incomplete, and someone is needed to finish off the work. This can lead to conflict and potentially delay completion of the PhD; good departments will have support mechanisms in place to arbitrate in such situations. The student may choose to agree on the basis that the work fits in with their own PhD direction and they can use the material in their thesis, or that there is payment at a reasonable rate, or the student may be well ahead in their own work and not mind doing something extra. In other situations, they should have the right to decline. For readers from an NHS background, it is noted that income generation by an individual is much more significant in the university than in the NHS, and this pressure may be passed from supervisor to student.

The relationship between student and supervisor is highly one sided: if the PhD goes wrong, resulting in the student failing their viva or not completing, then this is annoying or embarrassing for the supervisor but a career changing disaster for the student. The negative impact on the student of a poor student–supervisor relationship is probably more important than in any other staff–line-manager relationship. Consequently, more attention is paid to supporting and monitoring this interaction than to any other in the life of the academic. In a decent university, the PhD student should expect to be looked after very well, with well developed systems for monitoring academic progress and for supporting the student with personal issues.

1.4   What makes the PhD successful?

1.4.1   An ideal world

At the beginning of the novel Anna Karenena is the line 'Happy families are all alike, every unhappy family is unhappy in its own way'. A similar comparison can be made between successful and unsuccessful PhDs in that successful PhDs share many characteristics while unsuccessful PhDs are unsuccessful for a whole range of different reasons. Successful PhDs usually have several core features listed below:

- a student who is academically able, interested in the subject, willing to learn and adapt, determined to submit a thesis and willing to receive help and advice from others;

- a supervisor who is working at the cutting edge, has good communication skills, has the time to devote to supervision and has the student's best interests at heart;

- a thesis subject which is original and is in the field of expertise of the supervisor or supervisors;

- a department where there is interest in the topic of the PhD, which has PhD students working in similar areas to act as a peer-group, which has appropriate resources in terms of equipment, lab space, office space and computing;

- a university which has well-developed mechanisms for supporting students and good central resources such as libraries and sports facilities.

When a PhD goes wrong, it can lead to the student leaving of their own volition, not being allowed to continue after the first year, or even failing the viva (although this is very rare in most UK universities). It is also possible to pass the viva subject to extensive amendments to the thesis or even additional experimental work. Any student who worries that they may end up in this position should be aware of the support mechanisms available in their department and the wider university, and should feel confident that they will be taken seriously and assisted with their problems.

1.4.2 Student–supervisor and supervisor–supervisor issues

Assuming that the student is willing and able then by far the largest number of problems arise due to some form of communication breakdown. The most common communication breakdown arises from the supervisor who simply does not do sufficient supervision, especially in the early stages of the PhD when more support is needed. This situation is at its most extreme when there is only one supervisor. If there is a second supervisor, as is required in many universities, then they may identify the problem and act as a go-between, in some cases acting as the main supervisor. Lack of communication between the principal supervisor and the student is hard to solve and likely to lead to difficulties throughout the PhD. The student can get some idea of these issues at the interview by asking about procedures for supervision and by talking to other students in the group. If there is talk of little direction from supervisors and no regular supervision meetings then these should generate loud alarm bells in the head of the student being interviewed.

Multidisciplinary PhDs usually have a minimum of two supervisors and in fact the university will normally allow as many supervisors as needed, though more than four is unusual. Having more than one supervisor itself leads to communication issues. Supervisors, as academics do, will happily argue amongst themselves in front of the student who, until he or she becomes used to this form of academic discourse, may sit in the middle looking from one to another supervisor like a spectator watching a tennis match. Ideally the student should not receive contradictory advice from the supervision team: once all of the discussions have taken place there should be clear and straightforward advice to the student. Ideally there should be regular meetings of all supervisors and the student. These meetings should be the place where important decisions are taken about future direction. As with so many ideal scenarios this sometimes does not occur due to the pressure of other demands on the supervisors such as teaching, grant writing, administration, conferences and professional duties. Irregular meetings and contradictory advice can put the student in a difficult situation; which one of the supervisors do they take notice of? A choice has to be made and the student will usually choose one supervisor who he or she likes/trusts more than the others. When this is not the principal supervisor this can lead to tensions. Communication issues may increase dramatically with the number of supervisors, and in our experience, the ideal number of supervisors for a multidisciplinary PhD is two; three if the supervisors communicate extremely well with each other. Again, good universities will have robust mechanisms in place to recognise and address issues of this sort.

1.4.3   Interdisciplinary PhDs

PhDs at the interface between engineering/physics and biology/medicine may be divided into two phases: development and validation of methodology (software, hardware, experimental systems, biology and patient protocols) followed by experiments with a clinical or biological end point. For PhDs with a clinical theme, it is desirable to include some data from patients. If the PhD involves ground-up technology development then virtually all of the PhD may involve development of the methodology or instrumentation, with a few studies at the end showing initial patient data. If the core technology is already available, then methodology development is usually much quicker. For example, PhDs involving MRI are often equally divided between technique development and patient studies. The two phases of the PhD generally lead to two types of publication: papers on methodology development and validation which are appropriate for a technology journal, and papers with a biomedical focus which are more appropriate for a biomedical journal.

A student undertaking an interdisciplinary PhD may find themselves working in an unfamiliar environment, which can impact enormously on them. Long gone are the days when medical physics or medical engineering PhDs only took place in a department of medical physics and medical engineering where the student would be surrounded by like-minded people. Increasingly students with a background in physics and engineering are located in academic clinical departments with the student reporting directly to a medical first supervisor. Working directly with a medical practitioner represents an enormous opportunity for the student; to see at first hand the clinical issues which the thesis should be attempting to address and to have an interested and engaged clinician directing clinical aspects of the research.

These positives are balanced by the fact that most medics are not able to supervise technical aspects of the PhD. Students need appropriate technical supervision no matter where they are placed and the best scenario is where there is a medical supervisor and a technical supervisor (ideally equal first supervisors).

The other main issue related to location is publication and the dreaded impact factor of journals. For those readers for whom this is new, here is a brief summary. The impact factor is a measure of the number of times articles from a journal have been cited by other papers. Impact factors for the best journals in technical fields are typically 2–4 (e.g. Ultrasound in Medicine and Biology, Journal of Biomechanics) whereas the best medical journals have impact factors of 10–50+ (e.g. New England Journal of Medicine, Lancet). University research funding is linked to the quality of research output and there is an internal pressure on staff to publish in journals with a high impact factor relevant to the work of their department. If a student is located in Engineering or Medical Physics, then publication in a top engineering or medical physics journal is a sign of success. However, in clinical departments, an impact factor of less than 4 is not considered high impact. There can be directives from the Head of Department not to publish in journals with low impact factors, which include virtually all of the best engineering and medical physics journals. This scenario is relevant to many students today based in clinical departments and is a source of stress for both students and their supervisors. In our opinion, if there is no acceptance and encouragement for publication of the student's research in the top engineering and medical physics journals, then medical physics and bioengineering students should not be located in those departments.

1.5   Summary

The PhD is the main route for scientists to start their careers in research, as it has been for over 100 years. The successful PhD student will emerge a different person; one of an elite group of 1% of the (UK) population with a doctoral degree. There are significant challenges for the student, supervisors and universities in ensuring PhD success. These challenges are by no means stable and continue to evolve with the growth of interdisciplinary subjects such as medical physics and bioengineering. But most former students would agree that the personal and career opportunities that a PhD opens up justify the investment of time and effort.

2 Research associate and research fellow

Peter R. Hoskins and Stephen F. Keevil

This chapter will cover the period after the PhD studentship and before the salaried post (lecturer to professor).

2.1 Funding of research associates by grants

The term 'research associate' (RA) is often used to describe someone at this point in their career who is funded to undertake a particular piece of grant-funded work. Grants are time-limited, usually lasting 1–5 years with 3 years being most common. The grant when awarded is a form of contract between the funding body and the university in which the university commits to undertake a package of work. The grant application is led and managed by a full-time member of staff called the 'Principal Investigator' (PI). The RA is employed from funds obtained from the grant with the implication that when the grant ends, the funding for the RA post ends. The role of the PI and co-investigators is to deliver the funded research in full, on time and on budget. The role of the RA or RAs funded by the grant is to undertake the work for which the grant was awarded. This brief description emphasises the important difference between the PhD student and the RA. The RA has much less flexibility concerning the direction of their work. Whereas the PhD studentship belongs to the PhD student, the work of the RA is dictated by the grant application and the PI has responsibility and authority over research direction.

Research grants are the mechanism by which universities fund research. In broad terms, full-time members of staff are expected to bring in money through grant funding, and RAs do the work funded by the grants. Grant funding is from two main sources: government grant giving bodies (e.g. EPSRC – Engineering and Physical Sciences Research Council, MRC – Medical Research Council, BBSRC – Biotechnology and Biological Sciences Research Council, NIHR – National Institute for Health Research) and charities (e.g. Wellcome Trust, British Heart Foundation, Cancer Research UK). Generally, only established members of staff are allowed to apply for research grants. Fellowship grants are another issue and are discussed below. Grants are hard to get! Major UK grant-giving bodies have an overall grant success rate of 30%, with higher rates for larger strategic bids led by senior PIs and lower success rate for smaller project grants of 3–10%. The grant success rate of individual PIs can go up and down. Around 10% of PIs work in 'hot' areas, have good grant success over many years and maintain large groups. Most PIs have a more mixed success rate with groups which oscillate in size.

2.2   Career progression and planning

In this funding environment, the RA must be prepared to move institution to obtain their next post. Some RAs manage to work in the research group of one PI until they get a permanent academic position but this is not typical. It is much more common for RAs to move between PIs and to move between universities, sometimes involving major geographical relocation. Research is international, so this movement can involve going abroad for a period. Most academic departments have students, RAs and PIs from many different countries, emphasising the international nature of academic life.

For those RAs who wish to progress into a permanent academic post, it is best if they are aware of the expectations and culture of academic life so they can work out at an early stage if this is the lifestyle that they want. The early life of an academic, as a PhD student and RA, is dominated by hands-on research and the publication of first-author papers. The life of an established academic is dominated by writing grant applications, managing a research group, teaching and administrative duties. Very few senior academics have time to undertake hands-on research, leaving this to their group members.

In terms of progression in an academic career, around 30% of PhD students become RAs, around 3.5% will get a permanent academic position, and around 0.5% will become professors, noting that these numbers vary by institution and by subject area. Most RAs will therefore not gain a permanent academic position. They will at some point need to look for a job outside of academia: in industry, in the NHS or elsewhere, often completely changing career direction. It is therefore important that the RA acquires a set of skills which are relevant to a potential future employer. It is also important that the RA keeps an eye on their career direction, especially on questions such as 'where do I want to be in 5, 10, 15 years', 'am I likely to get a permanent academic post' and 'what do I do when the funding runs out'.

The skills and knowledge of a research group are embedded in the heads of the team of researchers. It is therefore in the PI's interest to employ a key RA for as long as possible. Some PIs will do this without much thought given to the RA's future career. Research associates can go from one short-term contract to another for many years with no hope of gaining a permanent academic position. This can lead to difficulties for the RA when the funding does eventually run out. Research associates may have developed highly specialist skills which have kept them employed in academia, but for which there are very few jobs elsewhere. Today, universities, and consequently PIs, are much more aware of their duties not to make their RAs unemployable and to make available appropriate career advice. Most universities operate mentoring schemes. Our advice is that all RAs should undertake career planning on an ongoing basis including making use of university resources, especially mentoring and transferable skills classes.

2.3 Metrics of success

Universities have been subject to a review of their research output since 1986; this was called the Research Assessment Exercise (RAE) until 2008, and the Research Excellence Framework (REF) in 2014. University core funding from central government to support, for example, the salaries of permanent academics (as opposed to grant-funded RAs) is largely based on the results of these reviews, to the extent that the RAE and now REF dominate the research agenda of all UK universities. Those universities with high REF results get enhanced funding and are able to expand their permanent staff. Those with poor REF results will get reduced funding requiring contraction of their permanent staff. The intention is to improve the quality of research in the UK, closing down low quality research and encouraging high quality research. It has been observed that, over time, this could lead to a division between teaching and research. Research may become concentrated in a few super-universities which compete at a world level. Other universities could, over time, become teaching-only institutions. To offset this, introduction of tuition fees for students in England and Wales, which is another very significant source of income for universities, has created more of a market in university education and raised the profile of teaching again.

The PI, and hence the RA, operate within this system and there is significant, often extreme, pressure to follow the guidelines of the next REF when it comes to publication of journal papers. The RA must undertake grant-funded technical work to a high level **and** publish the results in high quality journals. The consequences of not publishing papers are severe, especially for the final report which the PI writes for the funder when the grant ends. Increasingly, final reports are judged by simple metrics including the number of papers and the quality of the journal in which the papers are published. Poor reviews from final reports are not good for the PI and may affect the success rate of future grant applications. It has been argued that the RA has a duty to ensure that papers are written up as public money has been used to fund the research. These pressures are balanced by the reality of research where the results usually come near the end of the funding period, so that papers are being written up after the RA has left and moved on to another post/city/career. While it is fine to say that there is a duty to write up papers, the RA's legal obligations to the university end when their contract ends. In our experience, around one-third of all papers which arise from grant-funded research are written up after the grant has finished. In practice, most RAs will write up papers after they have moved on to the next post, albeit more slowly due to the pressures of their new job and with help from the previous PI. Others will not write up and the PI has to step in to ghost-write papers. This is time-consuming for the PI, takes the PI away from other tasks, and in some cases, is not possible as the documentation is insufficient or the work incomplete. In other cases, referees of a submitted paper require more work before publication, which may not be possible as there is no longer a resource (i.e. the RA) to do this. With the current university funding model, based on short-term contracts, inevitably some research will not be written up. Research associates need to be aware of the expectation to publish and of the duties on themselves to assist with this. Ideally, RAs should generate a steady stream of papers, with a completed piece of work followed immediately by a submitted paper.

2.4   Strategies for success

This discussion leads on to a related topic which is applicable to both PhD students and RAs, which concerns how the researcher goes about their work. Central to being effective as a researcher is being very clear about what the aim of the research is. To illustrate this, let us look at two approaches to research which look indistinguishable from the outside but which are vastly different in their effectiveness.

In the first approach, the researcher works in some general area but does not define a specific research aim. The researcher develops methodology, collects some data, refines the methodology, collects more data, and so on. The methodology becomes ever more complex and the data continues to accumulate. Meetings with the supervisor are characterised by the supervisor asking when the research will finish and be written up and the researcher insisting that the methodology needs refinement and more data needs to be acquired. Only after extreme pressure from the supervisor will a paper be written up, and often there is no paper output. This approach is not research, it is 'activity', where there is a continuous cycle between results and methodology.

In the second approach, there is a clearly defined aim. The researcher designs the methodology to address the aim, and collects the minimum amount of data to answer the research questions which arise from the aim. The paper is often written as the data are collected, sometimes before, with gaps for the results which are filled in as the research progresses. The paper has a clear aim and the conclusion is straightforward and related to the aim.

The authors' realisation that these two approaches exist (and variants between these extremes) arose from observations of different researchers over many years. These two approaches are in some ways indistinguishable: both involve complex methodology, collection of lots of data, and busy researchers. One approach is characterised by a clearly defined aim which by definition can then lead to a clearly defined conclusion and an end to the research. The other approach is defined by no aim and hence (also by definition) can never come to a conclusion or an end. One key message which is central to success in research comes from this discussion; make sure the aim is clearly defined and then configure everything to address the aim. The paper-led approach described here usually leads to a large number of first author papers and hence fulfils one of the main criteria required for progression in an academic career.

2.5 The principal investigator–research associate relationship

If an RA is to stand any chance of obtaining a permanent academic post, they must have first author publications which they have initiated. In other words, they must have evidence of being an independent researcher. However, as noted above, the RA is contracted to undertake the work of a grant which belongs to the PI, not to the RA. There is a natural tension in the RA–PI relationship due to the different needs of each. The PI is mainly concerned with delivering the grant, whereas the RA may wish to establish evidence of independence as a researcher. If not properly managed, this can lead to difficulties. If the PI is too controlling of the RA's workload, then the RA will not have time to develop their own research agenda. If the RA is entirely focused on their own research agenda, then the grant-funded work will not progress which is a nightmare scenario for the PI. Ideally, the PI is able to support the RA in developing their own research agenda. However, most universities do not have well developed systems for supporting development of RAs as independent researchers and there remains a need for universities and grant funders to better address this.

2.6 Research fellow

The issue of an independent research agenda becomes increasingly important for more senior RAs who are looking to move to full-time academic positions. The bridge to a permanent academic post is often a research fellowship. The RA (or indeed PhD student) can apply for a fellowship to fund their time for a research project which they have developed. Research fellowships are highly competitive and reviewers are looking for a very strong first author publication record, evidence of an independent research agenda and a very strong original project which fits with the funding priorities of the grant-giving body. The RA can apply to several grant-giving bodies in parallel in order to increase the likelihood of funding. If the fellowship proposal is outstanding, the RA may be offered fellowships from all of the bodies to which they have applied, though obviously only one can be taken up. Some fellowships are bound to the host university, others are transportable. Transportable fellowships obviously give the student more choice in terms of where they will be located. The savvy research fellow will use this as a bargaining tool in negotiation to get commitment from the university for a move to a permanent position once the fellowship has ended.

2.7 Interdisciplinary research

Much of what has been described so far applies to all RAs in science and engineering. Issues in research at the interface between physics/engineering and medicine/biology were described in Chapter 1 (PhD student). Many of those issues are also relevant to the RA, and mostly involve two aspects described below: line management and publication.

In research projects in interdisciplinary areas such as medical physics, medical engineering or bioengineering, the PI should organise regular project management meetings where decisions are made with regard to the direction of the work. These will include all grant holders and all researchers (RAs and PhD students). Ideally, the grant holders should have technical skills in all of the relevant areas of the project including those of the RA. This becomes important on large grants with several RAs working in different areas. At least one grant holder should be able to have a detailed technical discussion with the RA, be able to evaluate progress and to offer advice. Problems may arise if the work of the RA is outside the scope of expertise of the grant holders, who are then unable to make a judgement on progress or offer any technical advice. This can arise, for example, if a project is led by a clinician who does not have medical physics or engineering co-investigators to guide the technical aspects of the work. Generally only very experienced RAs, who are essentially operating independently and do not require supervision, can make progress when there are no technical grant holders. Potential RAs need to evaluate the mechanisms for adequate technical advice when applying for such posts. A related issue is the tendency in some clinical departments for physics and engineering researchers to be regarded as a support service for clinically-led research, rather than research leaders in their own right. This can lead to a lack of first author publications and may result in a lack of career progression for the researchers involved. In a good multidisciplinary department, the contributions of all members of the team to all aspects of research, including being PIs on technically-oriented grants, are recognised and encouraged.

The need for the RA to have a series of first author publications has been emphasised. If the RA is in a clinical centre, or on a grant led by a clinician, there can be pressure not to publish in the top medical physics and engineering journals as these have low impact factors compared to clinical journals. The opinion of the authors of this chapter is that, if there is no acceptance and encouragement for publication of the RA's research, as a first author, in the top engineering and medical physics journals then medical physics and bioengineering RAs should not be located in those departments.

2.8   Summary

The RA is a contract researcher hired to undertake and complete work funded by a grant. Successful RAs gear their work around publication of first author papers which is of vital importance for the grant and for the RA. Most RAs will not become permanent academics and need to maintain a constant watch on their career progression while acquiring transferable skills. Research associates who aspire to become permanent academics need to develop their own research agenda demonstrated through first author publication. The research fellowship is the main route to becoming a full-time academic.

3 Permanent member of staff (lecturer to professor)

Peter R. Hoskins and Stephen F. Keevil

3.1 Academic titles

This chapter describes the career of an academic from lecturer to professor. The terms 'academic staff' and 'academic' will be used as an abbreviation for 'academic member of staff'.

Some 3.5% of UK PhD students will become academic staff in universities. The traditional grades for academic staff are lecturer, senior lecturer, reader and professor. However, there are variations: in some universities, senior lecturer is effectively a higher grade of lecturer and there is also a principal lecturer grade. In other countries, and increasingly in the UK, titles such as 'assistant professor' or 'associate professor' are used to describe grades from lecturer to reader. These titles are used in inconsistent ways in different universities, which can lead to confusion.

3.2 The work of an academic

The work of an academic is divided between three main areas, corresponding to the three main functions of a university: i) research, ii) teaching and iii) commercialisation and knowledge exchange. The percentage of time spent in each area will vary during the career of an academic. Promotion (covered below) is dependent on achieving success in one or more of these areas.

Table 3.1 shows the 2014 income for Edinburgh University, a world top-20 university (www.thecompleteuniversityguide.co.uk/league-tables/rankings). The three major income streams in the table are associated with teaching and research, which historically have dominated the work of academics. New lecturers will (should!) have a very good track record in research. They will have much less experience in teaching (though some research fellow posts ramp up the teaching load during the fellowship). The new lecturer will be expected to contribute to teaching and in parallel develop their own research group.

3.3 Teaching

A new lecture of 1 hour duration typically requires 5–20 hours of preparation plus delivering the lecture and follow up with students. Historically, teaching was seen by many academics as being an annoyance, something which took them away from

Table 3.1 2014 income for Edinburgh University

Category	Income (£million)	Comment
Funding Council grants	204	Teaching support and research infrastructure
Tuition fees and contracts	194	Income related to number of students
Research grants and contracts	216	To undertake projects and fund studentships
Other operating income	147	For example, conferences, catering, donations
Endowment and bank interest	19	Interest on bank deposits and endowment

their research and which they did with the minimum of effort. Nowadays, considerable emphasis is rightly placed on the student experience. There is a consequent demand on academics to provide the student with high quality teaching. New lecturers, unfamiliar with the discipline of teaching, may be required to undertake courses in teaching theory and practice, including the option to sign up to gain a teaching qualification such as a postgraduate certificate. The experience of many academics with heavy teaching loads is that this dominates their life during term time and they struggle to devote adequate time to other activities such as research, administration and professional duties. The life of an academic involved in teaching follows the same pattern that they were familiar with as undergraduates. The beginning of the year is mid-September corresponding to the new intake of students. The end of the year is June when all exams are completed and marked. The completion of exam boards followed by the graduation ceremonies in July leads to an emptying of departments. Undergraduates go home, staff take their summer break. Departments which were busy with students and staff a few weeks before are now much less busy and some may be almost deserted. Committees shut down, emails reduce, staff have more time to attend to their research groups, and have time to write papers and grants. All feels calm for a few weeks then activity gradually increases as the new academic year approaches. There are exam resits, marking, exam boards and most importantly, students return on the first day of term in mid-September, and the new year begins again. This pattern is less clear in medical schools, where student vacations are much shorter and academics have less of a respite from teaching over the summer. Departments which host summer schools may also have a reduced break from teaching.

3.4   Research

The academic must initiate and sustain a research group. Academics are not provided with an annual budget from their university to carry out research. They have to gain funding to undertake research from grant giving bodies. New lecturers might begin their full-time academic career by doing their own research using shared equipment, or through collaboration. However, the expectation is that grant applications will be written which will be funded. As the research group grows with time, the academic has less and less time to undertake their own research (or decides that they do not want to do hands-on research anymore) so moves to managing a research group.

3.4.1   Grant funding

Academics must expect to spend a large amount of their time writing grant applications. The overall research council success rate for grants is around 30% with higher rates for strategic bids and lower rates of 3–10% for project grants. A typical project grant might take 2–3 man-months to work up. If the academic has a 10% success rate and does nothing but write grants for 2 years, they can expect to get one grant funded. This means across the UK an enormous amount of staff time is spent on writing grant applications which will be rejected. Academics tend to regard the grant funding system as being effective and well designed when they get their grants funded, and inefficient and outdated when their grants are rejected. Academics can choose between two strategies for grant funding. In the first, they have an area they are keen to work on, then look for funding opportunities. In the second, they see what grant giving bodies are focusing on and align their grant applications accordingly. Needless to say, success rates are much higher for the second approach.

An academic career typically lasts over 30 years. Entire research fields can come and go in this time, and the academic must be prepared to shift their research direction, even changing fields into new and emerging areas. Some academics with expertise in generic areas such as fluid mechanics can simply change their application area (e.g. from industrial to biomedical applications). This makes it relatively easy to supervise students and RAs in the new area as the physics and engineering are very similar. Changing fields to a completely new area is more difficult if it lies outside the technical expertise of the academic. This is a common feature of bioengineering research and a team-working approach involving collaboration with other academics is essential in order to provide supervision of all the key technical areas of a project.

3.4.2   Grant management

Many different skills are necessary for successful management of a research grant and of PhD students: project management, financial management, staff management, student supervision, just to name a few. Usually the PI of a grant is given considerable authority. The PI must operate within the rules of the grant giving body, and there are big differences in these rules between different bodies which the PI must be aware of. The PI must also operate within the rules of the university. It is the PI who has authority over staff working on a grant. The PI is accountable for the scientific delivery of the grant (in the form of the final report and associated journal papers) and for ensuring that there is no financial mismanagement. New academics are strongly advised to attend as many courses as they can on management relevant to grants and supervision of PhD students. Courses which assume everything behaves like clockwork from start to finish are of limited use. Good courses should provide advice on what happens when things go wrong. Further details on grant funding and project management of grants are provided in other chapters in this book.

3.4.3   Research timelines

In grant funded research, the time between initiation of a good idea and publishing the work can be many years. A grant application takes 3–18 months from idea to submission. Smaller project grants involving a single RA can be put together in a few months. Larger complex grants involving multiple partners and multiple RAs can take 12 months or longer. The review process is 6–9 months. If successful, it takes 2–3 months for the award letter to be released and an account to be created in the university from which the funds can be drawn. Advertising, interviewing, appointing, and waiting for RAs to work their notice period typically take 6 months, longer if a visa is required. The new RAs may need training if their background is in a different area and they will need to familiarise themselves with the work and get up to speed. The first papers appear 6–18 months after the start of the grant. All of this gives a long timeline of 2–3 years between idea and the first papers being submitted, longer for more complicated grants. This timeline can appear glacially slow to the new lecturer who, having undertaken their own research as an RA, will have had much faster times of 3–12 months between idea and paper submission. This slow process also means that the entire field has shifted considerably between idea and starting the work of a grant. What was a novel project when it was initiated may now be done by several groups in different parts of the world. In the authors' experience, these other groups are often much bigger and have much better funding! These delays are part of the current process of grant funding. There are grant schemes which are designed to short-circuit this glacially slow approach. While every researcher wants large amounts of flexible funding, these schemes are only open to a tiny minority of extremely successful PIs. These grants have a broad remit allowing the PI to respond more quickly to new opportunities than the standard project-grant will allow.

3.5   Academic grades and promotion

The various academic grades are described here in an abbreviated form with respect to research roles (these are the traditional grade titles; as discussed above, some universities use different titles).

- *Lecturer.* Has own independent research programme, involving own-account research, supervision of PhD student and first grants. Evidence of developing reputation in the field.

- *Senior Lecturer.* Has a research group with a developing track record of grant funding and publication. Evidence of national/international reputation in the field.

- *Reader.* Has an established research group and a track record of grant funding and publication. International reputation in the field.

- *Professor.* Has an established research group and a substantial track record of grant funding and publication. Clear evidence of leadership in the field at an international level and an international reputation.

In the past, promotion in the university was dependent on research activity with little account taken of teaching. The modern university recognises the contribution to teaching and most UK universities have promotion criteria based on teaching right up to professorial level. Due to word constraints, the remaining discussion below is with respect to the research academic path; teaching-track careers are considered in Chapter 20.

3.5.1   Metrics of research success

The definitions of academic grades above referred to grant funding, size of research group and reputation in the field. How are these to be evaluated in a manner which allows decisions to be made concerning promotion, and in a wider context concerning the success of an academic department? In recent years, a number of metrics have been used to define academic success. These are used in promotions and in research assessment exercises. A few popular metrics are described below.

- *Grant funding.* Research-intensive universities are extremely large businesses. Major UK universities have a turnover of around £1 billion per year, consequently those PIs that can bring in the money tend to progress in their careers. Some universities have guidelines for academics at an individual level for obtaining grant income, e.g. professor should bring in £150k per year. This can be challenging when there are wide variations in grant success. Other universities have criteria at the departmental level, meaning that a period of low grant success for the individual is not unduly problematic. In medical physics and bioengineering, a successful PI might have £600k of current grants equating to a group of two RAs and two PhD students. Stellar PIs can have huge income in the millions of pounds and research groups of 50+ researchers.

- *h-index.* This is an index of the impact of the researcher's publications in terms of the number of times their papers have been cited by other researchers. For example, an h-index of 20 means that the researcher has been an author on 20 papers each of which has been cited at least 20 times. The h-index increases throughout the career of an academic, and is highly field-dependent. In medicine, the h-index of a professor is typically 50+, in medical physics and bioengineering it is 20+, and in engineering 10+.

- *Number of peer-reviewed journal papers.* Historically, this had been the easiest metric to quote. A simple web of knowledge search reveals how many papers have been published. Professors in medical physics and bioengineering might be expected to have 50+ journal papers. In recent years, there has been less emphasis on this metric with the pressure to publish in journals with a high impact factor. However, for many academics, it is still an easy and relevant metric to quote.

3.5.2   Promotion strategies

When seeking promotion, the academic should consider their current and other institutions. In their current institution, the academic can put themselves forward through the promotion process which usually requires discussion with and support from the Head of Department. It can sometimes be useful to test the water by sending a draft application to the Head of Department or other senior colleague, as feedback is provided on what the academic needs to do to meet the promotion criteria, which then allows the academic to focus on these aspects before full submission. When moving institution, posts can be advertised and applied for in the normal way. It is also common for an academic to approach a university to discuss a possible move (or vice versa). In these circumstances, the key currency which the academic offers is the amount of grant money they can bring with them, especially if this is a good fit with the research interests of the new institution. When an academic moves, it is usual for most of their research team to move with the academic. This can be resisted by the host university as they lose equipment, grants and staff and usually there is a process of negotiation. Some grant giving bodies assign the grant to the PI who has the right to take the entire grant to the new institution.

When the academic has a transportable fellowship (which can be awarded to fixed-term and permanent members of staff) or has just landed a large grant, these are the times when the academic should look to see where they want to be located longer term. Does the current institution offer the right career opportunities or should the academic consider moving elsewhere? Is the current city where the academic wants to live long-term or do they want to live elsewhere? This process of moving research groups between universities happens all the time and the academic should feel no guilt in using this method to their advantage.

3.6   Later stages of academic life

Sustaining a research group over many decades can be challenging. Rejection of several grants in a row is followed by shrinkage of the group and loss of key expertise, in some cases, ending particular avenues of research. Universities offer sabbaticals, where the academic spends time in another lab in the UK or abroad. This can help the academic gain new skills and new collaborations which can then be used in grant applications. Long-term lack of grant success will lead to a reduction in publications. The various metrics against which the academic is judged in research assessment exercises will not be met. Universities do go through periods of shrinkage and the academic without a research group may volunteer for severance, opt for early retirement or in some cases, may be made redundant. Many academics judge that a research-only post is more high risk than a combined teaching-research post. When grant funding goes down, the academic can opt to take on more teaching duties.

The final paid stage of the academic career is that leading up to retirement. Academics will have built up experience and skills over a lifetime. Many academics use the last 2–3 years of their paid working life to hand on to the younger generation, assisting collaborators and members of their research group in grants that they will lead. Following retirement, some academics stay on in an unpaid associate role (for professors, this is called 'Emeritus Professor'). They maintain access to library facilities and may have a desk to work at. Some retired academics continue to write papers and be involved as part of another academic's research group. Others use the university more for pastoral purposes, maintaining contact with colleagues.

3.7   Summary

This brief chapter has described the career of the academic staff member. It has mainly discussed the research career with a lesser emphasis on teaching. The academic needs to be mindful of the various metrics against which their performance is evaluated, and which are relevant for promotion and continued tenure. The academic must be prepared to change field to take advantage of new opportunities, and to change institution when the opportunity arises. Combined teaching-research positions are less risky than research-only positions, but come with challenges of overload during key teaching times. Many academics never really retire, they just stop receiving a salary.

4   Academic leadership and management

Michael A. Smith

4.1   Introduction

Leadership and management are important at all levels in all organisations of whatever shape and size, and this chapter considers some key aspects relevant to an academic environment. As we will see, it is important to consider this, not only in the context of a department or research group, but in the wider context of the whole institution, with consideration as to how professionals such as ourselves can engage in leadership and management activity at all levels.

At the outset, it is helpful to have some clarity around definitions, specifically the three key terms which will be mentioned and expanded on in this chapter:

- Leadership. Influencing behaviour (of an individual, team or organisation) for the benefit of the organisation.

- Management. The process of working with and through others to achieve the goals of the organisation.

- Administration. The organisation, procedures and tasks which are needed to support the operation of an organisation.

In this chapter, I will be mostly talking about leadership and management in an academic environment. I have deliberately avoided writing yet another short guide to general leadership and management principles, which are already abundant and can be found elsewhere. Instead, I will be describing some key aspects of leadership and management and specifically development opportunities. The final part of the chapter, however, will contain some suggested reading that would support personal leadership and management development.

I assume my audience will be medical physicists, medical engineers, biomedical engineers and perhaps other scientists who are pursuing a career which may be in a university or which may be in the health sector and associated with a university. I hope the insights I'm able to offer enable fellow professionals to identify leadership and management opportunities that will enhance their career paths.

4.2   Academic organisation and structures

The nature of academic organisations has evolved considerably over the past 25 years with a variety of different structures now existing across the university sector. The nature and roles of what we could regard as 'traditional' academic speciality-based departments have changed in many organisations whilst there has been a growth in multidisciplinary groups or centres, which provide the impetus for academic developments across subject boundaries. In addition, there has also generally been a move towards a small number of large faculties or colleges with varying degrees of devolved responsibility, both academically and managerially.

For a medical physicist, medical engineer or biomedical engineer pursuing an academic career, it is important to understand how the university in which, or with which, you are working is organised and how it functions in practice. This is important to be successful in the job you are currently pursuing, as well as offering opportunities for career developments to more senior positions. It means understanding the organisation structure, how and where decisions are made, and the processes you will need to comply with to be successful. Unfortunately, there is not a common model across universities so you will need to consider and analyse your local circumstances.

Most organisations, and universities are no exception, will tend to have an 'organogram'; this is an organisational chart in the form of a diagram that shows the structure of an organisation and the relationships and relative ranks of its parts and positions/jobs, indicating how the organisation is organised and managed, with lines of responsibility indicating the decision making process. Many such organograms often contain both solid and dotted lines; these need to be viewed with caution as they do not necessarily offer a clear explanation of the lines of responsibility in the organisation.

I have always found it important to consider the organisational structure of an institution in terms of 'form' and 'function'. Simply put, the boxes that describe the different components of an organisation in an organogram, such as a faculty, departments, research groups, central support units, represent the 'form' of the organisation. They are the building blocks which, when put together, form the complete institution. So far so good, but that is only part of the story. Of greater importance is the analysis of the 'functions' and the consideration of how the different components interact and what specific functions lie where. Examples of 'functions' range from financial planning and budgetary control and the approval for the appointment of staff at one end of the spectrum through to the authorisation of travel expenses and the acquisition of a larger filing cabinet at the other end.

Put at its simplest level, as an academic who may wish to undertake a new research programme or new teaching course, how do you manage what you do? Where is approval or authorisation given, where does financial planning responsibility lie, where might budget approval lie, which parts of the organisation have responsibility for giving you support, what should you legitimately expect from others and what are their expectations of you? And then if things are not working effectively, or you are faced with what you think are unreasonable challenges and behaviours (not an uncommon occurrence in a university setting), to whom can you go to resolve the problem or where can you go for support to help you resolve the problem yourself?

Probably the first step on the ladder of academic leadership and management is your analysis of your own institution, initially in terms of the form of the organisation but then, more importantly, your analysis and understanding of the multiple functions that enable the organisation to work and which you will need to work for you. Advice from more experienced colleagues is always valuable, ideally triangulated by comparable advice from a second source.

One feature of the organisational structure of a university that is worth highlighting is the potential difference between more traditional structures, such as departments, and newer structures such as multidisciplinary research groups/centres. Generally speaking (and I know there are exceptions), the former often has a greater degree of control and influence because the devolved responsibility for finance, staff and other resources rests there. The multidisciplinary research group/centre may be set up as a 'virtual centre', and though it may have a much higher external profile, it may have a relatively small budget and little control over its staffing. Such multidisciplinary research group/centres in the medical and health area are quite common.

Another feature of being an academic medical physicist, medical engineer, or biomedical engineer is that formal links may exist with the NHS. This not only creates a situation where we may be considered as having divided loyalties, but also means we may become engaged in activities which are not considered core to a university. This also applies to comparable staff in NHS departments who hold honorary university positions. Both the NHS and university will be aware of potential divided loyalties, and both may consider that the activities of the other are not part of their core business.

4.3   Two cultures

A factor which can be relevant to medical physicists, medical engineers, or biomedical engineers is that we can also be registered professionals in the medical and health sector, as well as academic specialists, enabling us to work in both NHS and universities. It would be prudent therefore, in the context of leadership and management, to consider how we should respond to the demands and requirements of two separate organisations and how we respond to the biblical advice that 'no one should serve two masters...'.

As professionals, we inhabit a world of research, innovation, education/training, and scientific application, which can exist in either the NHS or a university environment. It is not uncommon for members of our professions to be employed by either the NHS or a university, and hold an honorary appointment with the other organisation. Despite NHS hospitals having specific long-term partnership arrangements with universities, generally in the form of teaching hospitals, such appointments can sit uneasily with formal, and indeed employment law, issues. Essentially if a person is being employed by, and therefore contracted to, one employer, then there can be the implicit understanding that an individual's primary responsibilities, and loyalties, should lie with that institution. This may be the situation in practice, despite clear understandings at a corporate level that an individual's post may be split 50/50 between the two organisations.

Because we inhabit these two worlds, each with their own culture, processes and organisational structures, it is important to understand the inherent differences between the two organisations in order to exhibit and utilise effective leadership and management skills.

So, let's start first of all, with the NHS. Those of us who have worked in the NHS for many decades have become acclimatised to continual change, restructuring and reorganisation which occurs on a regular basis, often as a result of political imperatives. This chapter is not going to spend time describing the current organisational framework, but instead reflects on some fundamental aspects of the NHS, which remain constant, regardless of the organisational structure currently in existence.

Specifically, the NHS has had a management culture for far longer than universities have. The profession of 'hospital manager' has been around for about 50 years, and back in the 1970s, I can remember a relatively small team of managers providing the managerial input into running a hospital. However, as a young physicist in Edinburgh, I observed that the hospital managers had direct input into many departments, with the exception of science-based departments, such as medical physics and engineering, where the professional scientists would assume the managerial roles in the department. I certainly didn't observe this happening in clinical departments where consultants operated in a more individual manner. Thus, I was aware that our hospital-based profession offered the opportunity to develop management skills and utilise them in practice, probably more so than in other specialties. Departments would often manage the provision of scientific services, technical support services, the creation of new developments and the implementation of innovations, in a way that was somewhat different from the general hospital environment.

So you may ask, what's this got to do with academic leadership? Well, it has always been my view that management and leadership go hand-in-hand, and if you are to become an effective leader, whether it be academic or professional, it is important that leadership attributes are underpinned by effective skills and experience in management. At the very least, a practical understanding about what is possible and what is achievable, and how something may be implemented, is important in establishing a leadership relationship with others. So the environment in which many of us find ourselves, for example working at the interface between the NHS and a university, provides us with the opportunity to develop and hone our management skills in a way that may not be available for other specialist medical and health professionals.

It is interesting to consider what happened when there was a new initiative in the NHS that required leadership to take it forward. In the 1990s, for the first time in its history, the NHS started to develop an R&D strategy. Quite suddenly, hospitals were considering appointing directors of R&D to develop strategies and new processes and then implement change. Interestingly, quite a high proportion of those hospital R&D directors were medical physicists, certainly a disproportionately high number given the relative prevalence of our specialty. I would suggest that this occurred because, as a profession, we had developed skills in management, and attributes which allowed us to apply these in a new area and take on a leadership role.

The NHS, one of the largest organisations in the world, is very clearly part of the public sector and so at times will experience levels of direct government intervention. Universities, on the other hand, are autonomous organisations, with their own governing bodies, and most importantly, are not part of the public sector. This is a surprise to many people, particularly as they have many attributes and behavioural characteristics that would not be out of place in the public sector. Thus, there are similarities, enabling skills and experience to be transferable between NHS and university organisations, but there are also differences which must be understood if one wishes to take on leadership roles.

I mentioned earlier the culture of management in the NHS. At the time I was becoming aware of this, I also noticed that universities had a far stronger culture of administration as opposed to management. I observed the strengths in universities as they administered large-scale and complex processes; these strengths were different from the management strengths in the NHS that were generally associated with change management and the response to the growing and continuing financial constraints.

4.4 Opportunities for personal leadership and management development

We can consider a simple definition of leadership as: (i) identifying the direction to go, (ii) identifying a means of getting there, (iii) having the ability to motivate and support others to participate on the journey, and finally (iv) ensuring that any rewards are fairly shared on completion. Research therefore, offers a valuable and practical way into leadership development for young physical and engineering scientists. A university academic environment will not only offer the opportunity to instigate new research activity, but also allow an individual to assume responsibility for generating the financial resources required to undertake that research, and then project manage the research programme. A researcher can effectively serve a high level apprenticeship as a post-doc and research fellow before becoming a principal investigator, with the personal responsibility that that entitles. This offers individuals the opportunity to develop substantial management and leadership skills whilst undertaking a research project or research programme. Staff can be encouraged to pursue this path, but it very much requires self-determination and the willingness to acquire new skills and knowledge to be successful.

The acquisition of funding for a programme of research will require an academic to develop good project management skills, time management skills, communication skills and financial skills if the funding application is to be credible and successful. Universities will have research support units to assist in the preparation of funding applications, but these generally concentrate on the fine detail, and require the academic to have an overview and understanding of all the management aspects of the programme as well as the research component. Later chapters in this book deal with grant funding and project management.

Managing a research grant will certainly include the management of resources and the responsibility for ensuring that the research programme is costed properly and then remains on budget. As important, and possibly more so, is the experience that managing a research grant offers in recruiting and managing staff. In my experience, it is the recruitment and subsequent management of staff during a research programme that offers the first exposure to leadership as well as management. The tasks of identifying the type of person you wish to employ, the implementation of the current, generally fairly complex, procedures associated with recruitment, and then deciding whom to select, are all very important to master. Gradually, it will help you to understand the important balance and interaction between formal processes and personal judgement.

Having made an appointment, that person will then see themselves as working for you, whilst being part of a large department. Such a relationship enables the grant holder to begin to develop leadership skills in areas such as motivating staff, teambuilding and conflict resolution, though hopefully with not too much practical experience of the latter. This environment will also enable the grant holder to learn the importance of both managing up (building a good relationship with your own line manager to get the best results for you and your organisation) and managing down (building a good relationship with your subordinates to get the best results for them and your project).

In terms of personal leadership and management development, there is going to be the question of what you should do to maximise your chances for career development. Clearly there are a number of key 'transitions' in an academic career for which you would need to demonstrate different levels of experience and competencies. Taking on a variety of roles and assuming management responsibilities will gradually expand your experience and deepen your competencies. However, I would be cautious about thinking about serving on committees as a route to promotion. Most certainly take on your fair share of committee work, but consider what personal development may be achieved by your participation. Most importantly, think which central committees offer you opportunities to demonstrate your wisdom/judgement and attention to detail, and then present yourself as a possible member. It's always best to be proactive (most of the key committees seek staff nominations) and select what you do, rather than wait to be approached to serve on a committee that does not appeal or offer sufficiently useful personal opportunities. However, be aware that membership of many committees may largely enable you to develop your administrative skills and understanding of processes, rather than helping develop your skills in leadership and management.

We are aware of the importance of the research excellence framework to universities, and as a consequence, the way that leadership and management skills in research can lead to senior positions in the academic sector. This is, of course, also true for teaching. For readers who are employed by and work in the NHS, it is important to stress the huge impact that the research assessment exercise (RAE) and the more recent research excellence framework (REF) has had on the university system. Given that research and development activity has always been seen to be a key component of the work of physical and engineering scientists in both the NHS and the university system from the start of our profession, the focus on a quantitative assessment of research quality has increasingly differentiated the work of physical and engineering scientists based in the NHS and in universities. I would argue that, rather than cementing links between the NHS and universities, the RAE and REF have both highlighted differences in the quality of research undertaken in the NHS and universities, and also made it harder for a physical or engineering scientist to be both a successful academic and a successful practising NHS clinical scientist at the highest level.

4.5   Approaches to leadership and management

It is worth starting with working definitions of leadership and management and understanding the differences between the two. Leadership is a broader concept than management and can be considered as being about influencing the behaviour of an individual or a team, and perhaps ultimately, in a university context, a whole organisation. Management is about working with and through others to achieve the goals of the organisation. Management therefore focuses on systems, procedures, policies, structure, and control with an emphasis on the status quo. Leadership, on the other hand, is about what we are there to do and why, the effectiveness of activities and the creation of a vision for others. It is also about creating trust, inspiring others and motivating them to achieve clear goals that they themselves set.

Leadership in an academic context is about creating an environment for enterprise (i.e. the exploring of new ideas), bold moves and imagination, in the areas of teaching, research and professional practice. Essentially, academic leadership is not different from leadership in other organisations; what academics want is an academically inclined version of what people in most organisations want. In a departmental or research group context, this means having clear goals, a climate of respect, and co-operative authority structures that provide optimal conditions for professional activity and productivity. A paradox of academic leadership is that sharing power by empowering colleagues increases one's authority as a leader (Ramsden, 2003).

It is interesting to consider what individuals and groups want from leaders. Research has shown that the key components that people want from someone in a leadership role are: (i) purpose, meaning and direction, (ii) capability to engender trust, (iii) a positive attitude and (iv) a bias towards action and achievement (Bennis and Nanus, 2003).

During the last 10 years of my career, I have become increasingly aware of the crucial importance of trust in the context of leadership in particular. That's not to say that I wasn't previously aware of the importance of trust in my profession life; however, I was not aware of the complexities of trust and the need to place an overt emphasis on it. I had previously observed situations where individuals or organisations did, or did not, trust each other with the result that trust as an attribute was treated like an 'on/off' switch. I would suggest that, as an academic leader, it is important to consider: (i) how to develop trust, (ii) how to maintain and grow trust and (iii) how to regain trust when it has been lost. Aphorisms about trust abound (if you google 'trust quotes' you'll get 130 million results) and though there may be a kernel of truth in many of these, it is better to adopt a strategic and tactical approach to trust (Flores and Solomon, 1998).

Trust can be built, often as part of a 'trading' relationship that can demonstrate the ability to do what has been agreed. At an organisation level, this can then lead on to productive partnership working and 'co-creation' (see Chapter 18 on **Innovation and knowledge transfer**). It is when agreed things are not done, or information is not shared, that trust starts to waver. An open communication approach, a 'no surprises' policy, and a clear response in terms of pro-active mitigating action are all essential. If something agreed has not been done, avoid excuses, and concentrate on taking mitigating action; this can mend trust if it begins to waver.

The importance of a bias towards action and achievement is also worth expanding upon. It has been shown that what people want in a leader is someone who: (i) gets things done, (ii) can convert aims and goals into actions, (iii) has the personal drive to ensure the goals are met, (iv) sees things through to completion without being distracted by new opportunities, and (v) is able to take difficult decisions. To operate effectively in all these areas, there is a need to be an effective manager and adopt an approach of 'action centred leadership' (Adair, 2009). This proposes that a leader must focus on three key elements: (i) achieving the task, (ii) managing the team or group and (iii) managing individuals. This approach very helpfully outlines the leader's responsibilities as a manager in each of these three areas, in order to ensure the balanced approach necessary.

The last aspect of leadership and management I want to cover is that of *transition* which deals with an individual's progression into management, or through the layers of an organisation to ever-higher levels of management and leadership. One of the main problems of progression is that the critical career passages between levels of management are not addressed by either the individual or the organisation. This can result in the potential for the individual to be 'derailed' later in their career, and may also be responsible for the fact that an estimated 50% of managers do not work to their full potential. It is important not to assume that 'what made individuals successful in the past will guarantee success in the future' (Handy, 2007).

The first critical transition is from managing yourself to managing others; in the case of academia, this can mean becoming head of department or head of a research group or teaching team. The second transition is probably the most challenging transition, and is from managing others to managing managers; in academia, this can mean the transition from head of department or research group to Dean of a faculty or school, often with executive responsibility. This transition is quite different from the first and requires different skills. The role becomes more strategic and focuses on developing plans for the future and networking within the university becomes more important. A common failing in this transition is when an individual tries to run a large faculty in the same way that they would run a small department or research group, without the institution or the individual recognising the different skills needed. The third critical transition is from managing managers to managing functional experts, i.e. the transition from Dean to becoming a Pro-Vice Chancellor, a role which is even more strategic and where the focus on developing plans for the future is critical (Jones; www.advancedlearningassociates.co.uk).

I hope the previous paragraphs strike a chord with aspiring academic leaders. The challenge for any of us then, is how we can use our own personal skills and strengths to achieve these outcomes. What is important to recognise is that, as individuals, we have different inherent strengths that we utilise in leadership and management roles. A very useful exercise is to explore and identify your own strengths and, as a consequence, identify the areas where you may have less inherent capability and therefore a need for personal development. I would suggest a publication which is linked to an online assessment website that can provide this information very effectively (Rath, 2007).

Finally, I'd like to mention that, in the UK, we have a Leadership Foundation in Higher Education, which is an independent body, supporting individual and organisational leadership development in the Higher Education sector (www.lfhe.ac.uk). They offer many courses covering a range of subjects, which are available to academic staff.

29

4.6   Recommendations

- Recognise that, to achieve a senior position in the academic sector, you will need to develop strengths in leadership and management as well as achieve success in the academic activities of teaching and research.

- Understand the difference between administration, management and leadership; there are many excellent texts and courses on the subject. Assess your own strengths and be aware of your weaknesses and how you can address them.

- Understand how your university is organised in terms of form and function and develop an understanding of what you can expect others to do, and what you will need to do yourself. In practice, you may have to assume more responsibility for some management aspects of your work than you would have wished or expected, due to weaknesses in university systems and processes.

- Take on responsibilities that will expose you to the different components of management and leadership, being clear in your own mind what you expect to gain. Be pro-active and make sure you take on any responsibility for a specific time-limited period.

- Take responsibility for your own personal development and ask to attend courses on leadership and management, both internally and externally organised; your university will have a budget for staff development.

- Seek coaching or mentoring in specific aspects of management or leadership at the relevant stages of your career.

References

Adair, J.E. (2009) *Effective Leadership. How to Be a Successful Leader*, 2nd edition. Pan Books.

Bennis, W. and Nanus, B. (2003) *Leaders: Strategies for Taking Charge*, 2nd edition. HarperBusiness Essentials.

Flores, F. and Solomon, R.C. (1998) Creating trust. *Business Ethics Quarterly*, Vol 8, Issue 2, 205–232.

Handy, C. (2007) *Understanding Organisations*. Penguin.

Ramsden, P. (2003) *Learning to Teach in Higher Education*. RoutledgeFalmer, Abingdon, Oxon.

Rath, T. (2007) *StrengthsFinder 2.0*. Gallup Press, New York.

5 Later life working and career opportunities

Michael A. Smith

5.1 Introduction

Much advice is proffered to assist young professionals at the start of their career, and subsequently to help them develop skills. However, this chapter is written for the other end of the age spectrum, specifically for those who are at, or moving towards, the traditional end of their career. In the last 20 years, there have been a number of substantial changes that have had an impact on the latter stages of our careers, whilst other changes offer opportunities that may have been more difficult to realise in the past.

The obvious change has been the removal of the age of retirement, giving individuals the opportunity to continue working, should they so choose, and remain actively involved in their work late into life. The removal of the enforced retirement age therefore makes it much easier for someone in later life to continue working. Should you choose to continue working, I suggest it will be important to address any concerns that employers may have, particularly around capability, if you wish to retain employment in a senior post with significant responsibilities. This should be done in a proactive manner so that an employer is assured that age will not be a risk factor and that processes are in place to identify potential problems before they occur. There will undoubtedly be a point where one's age will prevent one from performing sufficiently well at a senior level; retirement from such a position before that time occurs, is important for both the individual and employer.

Another factor, which is more insidious, has been the changes in the working environment which gradually, and over quite a long period of time, can contribute to an individual's disenchantment with their job and the way in which they are required to work. Earlier in my career, I cannot think of any examples of older professional colleagues taking the decision to retire early, unless for clear health reasons. Nowadays, it is much more common for someone to take a decision to retire early, in many cases because of dissatisfaction and disaffection with how their job has evolved, and how they have to work as a professional.

Specific factors, which I have observed contributing to such disaffection, are many and various. The first and obvious one is financial pressure. The continued drive to 'do more for less' in the health sector is always a challenge, but to do it year-on-year, decade after decade, with no end in sight will have an obvious impact on a person's resilience in their job. The second factor is the increasing audit and compliance culture, which has resulted in a multitude of processes and boxes to tick; not only is this time-consuming and somewhat wearying, but it also essentially implies a lack of trust between an employing organisation and an individual professional. I fully accept the reasons and rationale for audit and compliance, both in terms of assuring the necessary protection for the organisation and for patients, but the unintended consequence in my experience, is that increasingly, professional staff become disaffected over time.

The third factor is what I have come to call 'impression management'. Not only do we have to work effectively and perform well as professionals, and I presume none of us would have any problems about being judged against those criteria, we also have to spend increasing amounts of time managing the impression of success. This is best demonstrated by the desire for examples of success, good news stories, etc, which are continually being used, as well as some of the academic social media sites which emphasise personal 'impact'. The increased emphasis on the impression of success as opposed to more thoughtful, longer term, analysis of real impact is both time-consuming and far less satisfying.

These changes are in addition to another underlying factor which may be present; this is the fact that many of us start our careers as physicists or engineers working in medicine and, as we grow older and progress through our profession, whether it be in the NHS or university, we become increasingly involved in management, administration and politics. As a consequence, we spend less and less time doing the things which interested us and attracted us to our career in the first place.

Set against this range of negative factors, are the facts that not only has life expectancy increased, the age until which individuals remain active has increased substantially. We now have the opportunity, health permitting, to consider whether and how we want to work in later life, with the prospect of refocusing our activities, or indeed moving into new areas.

5.2   Seeing later-life as a career opportunity

I would propose that it is worth separating the concept of retirement from the desire to step off the treadmill of full-time working for a single large employer. It is possible to work in a different way so that the problems, challenges and time pressures can be removed, to be replaced by a work schedule that matches your personal interests and requirements, taking you into new areas with different ways of working, and new people to work with. There will of course be new problems, new challenges and new time pressures but if you approach it properly, you can retain greater control of them. And most importantly, you can have the option of saying no, and doing something else instead.

The move from so-called full-time work to part-time work is quite interesting if you deconstruct it. For many of us with a senior executive position in a large organisation, a full-time job can mean working perhaps 60 hours a week, sometimes more, when you include internet work time. This was certainly the case for me, so when

I was 60, I fulfilled a long-term intention and resigned from my full-time job, reduced the hours I worked, and started a portfolio career which allows me to concentrate on activities I enjoy. So now it feels that I only work part-time, that is about 35 hours a week, compared to my previous employment. I refer to this as 'pseudo-retirement', and I can envisage me being able to work, health permitting, for a further 15 years. The flexibility and personal control that this arrangement offers, are likely to extend my productive working life, and the enjoyment I receive from it, for many years longer than would have been the case if I hadn't set up a later life work strategy when I was young enough to reap the benefits.

5.3   Organise yourself

There are a number of ways that you can proceed. You can simply continue to be employed by an organisation in a new capacity, though you are likely to have to organise it yourself. This has attractions in that once you have negotiated the contract arrangements, they will simply pay you and the tax implications will be straightforward. It is not uncommon for a large organisation to discuss retirement and offer a 1- or 2-year contract, for perhaps a couple of days a week, after retirement. Should you proceed down that route, you should understand that your employer can decide not to renew your contract. You need to be aware that (i) your employer may be adopting this approach to ensure that you do retire, with a view to stopping the short-term contract in the future and (ii) the delay in setting up a new way of working may make it harder for you to do as you get older.

Setting up your own company is an effective way of proceeding and one that I have found helpful. In making this suggestion, I'm not discussing and certainly not condoning, ways of working which minimise tax. Much more important is to create an environment which allows you to work effectively, helps you secure future work and gives assurance to HMRC. You can set yourself up as a registered company and channel all your work through the company. I myself set up a limited liability partnership (LLP), which is a form of company, but one that allows groups of individuals to work together. It appeared to me to be an ideal vehicle for people interested in working later in life, offering as it does the benefits of individual freedom and control, cost sharing and the potential of working with colleagues on projects. Not only does it allow me to manage my work activity, but I can invite others to be members of the LLP, and increase the work opportunities.

One advantage of a registered company is that all of the tax issues are dealt with in a way that is right and proper. More importantly, however, is that some activities are pursued more effectively through a company. Some organisations and companies have a preference for dealing with registered companies rather than individuals who are working as sole traders. Their own internal processes make it much more complex to pay an individual rather than paying an invoice from a company; the complexity of their own internal processes actually deters them from doing business with individuals. Also there are some funding bodies which will only make payments to organisations, registered companies or registered charities, but will not make payments to individuals. Thus, if you have a company you can access these funding sources.

In my own later-life working, one area in which I work is helping organisations access funding which is associated with innovation and research; having a registered company ensures that I can receive payment directly from the funding body, thus mitigating the risk of not receiving funding promised. My strategy in this area was influenced by past experience, when I observed large organisations retaining money they should have passed on to their partners, who were working as individuals, and who had helped them prepare an application.

A final advantage of having a company is that it does enable you to differentiate your later-life work environment from your personal activities. One important psychological factor may be that, when you had a workplace, it was more obvious when you were working and when not. The absence of a specific work office environment means that other methods may be needed to distinguish work from non-work. It may also be that when working, we all need some triggers to engage the brain in a particular way, and also triggers to disengage, particularly if working from a home environment. I certainly feel more comfortable being able to compartmentalise my activities; perhaps this is just a function of age and the desire to minimise the efforts associated with multitasking.

Finally, whether you set up a company or not, many of you would have been used to having a support network when you were working for a single employer. When you move away from this environment, suddenly there is no one to make appointments, arrange travel, fix your IT, file, and generally make life more bearable. My advice is to find someone that you or your company can employ for perhaps a day a week, once you have started to generate an income. The number of people who can provide such support activity is increasing, most working from home, and utilising internet communication and shared access to e-mail accounts.

5.4   *Pro bono* work

Continuing to working later in life allows you to take decisions about how much *pro bono* work you may wish to undertake. You may be deluged with requests from lots of organisations to sit on committees or boards to give advice, for which you will not get paid; these can come from a range of organisations including professional, university, the NHS, charities and also private companies. To some extent, this is because, whilst we were working for a single large organisation, we would often sit on committees without any payment being received, the assumption being that this was part of our job and that our employer would bear the cost. When you think about it, this is rather a curious situation. Though there is an element of mutual dependency between some organisations, I was always aware that some organisations seemed to benefit far more out of this arrangement than others.

The term *pro bono* is actually a shortened version of *pro bono publico,* which is Latin for 'voluntary for the public good'. So should you decide to undertake *pro bono* work, I suggest you may wish to give some thought as to whether it is for the public good, and whether the organisation asking you should, in all fairness, pay you for your work because it is actually for their good and they can afford it.

One way to undertake *pro bono* work effectively is to become a trustee of an established charity, and once you become a trustee you are likely to be very quickly asked to become a member of one of their subcommittees such as finance. If you are interested, then simply approach the chair of the board of trustees or the chief executive of the charity in which you're interested. Alternatively there are a number of small head-hunter companies that specialise in finding trustees for charities. If you have not been a trustee before, it is worth attending a short course in order to understand the requirements and responsibilities of being a trustee; often such courses are run by the larger charities.

5.5   Getting paid

In your later-life working, you are now in a position to decide who deserves to access your wisdom and knowledge for nothing. I learnt over many years that the attention and thought that people give to any advice they have received is dependent on how much they have paid for it. When people get advice for nothing, they generally happily ignore it if it doesn't suit them. When people have paid money for advice, they feel obliged to give it serious consideration. And if they have paid a lot of money for it, then not only do they give the advice very serious consideration, they generally take it. As a consequence, I make no apologies for charging for my time and for my knowledge and advice, whilst devoting a proportion of my work time to *pro bono* work for organisations which I believe deserve it.

You will need to work out how much you are worth. Work out your fee rate and be very open and upfront about what you charge. I myself have a range of charges which I offer to prospective clients. I have a specific day-rate which may be reduced if the contract is for a sufficiently large number of days. Also if I'm asked to undertake a task, for example, doing some work in a foreign country for a week or so, the contract will include payment for a comparable number of preparatory days.

Generally, you will need to be flexible if you want to be successful with some contracts, though that does not necessarily mean reducing your price. Remember that if you reduce your price once, then you are very unlikely to be able to increase it in the future for that organisation. Sometimes a client will be restricted by the amount per day they are able to pay you. That does not have to be a problem provided the day rate multiplied by the number of days they are willing to pay for, matches your requirements.

I also have a reduced fee rate which is used for selected clients whose limited ability to pay is genuine and where I wish to undertake the work. Some of the work I do for the NHS falls into this category and I do it because I have a soft spot for the NHS and I wish to work with the individuals on the specific innovative projects. I also use the reduced fee rate in selected countries.

Lastly in this section, make sure you get paid. You will need to send invoices, and often chase invoices if you want to receive payment for work that you've done. Also do not let the amount you are to be paid get too large before you send an invoice. This approach also mitigates your risk. Since starting later-life working, I have experienced quite large organisations reneging on agreements to pay for work undertaken; finding this out early is important. This is not that unusual in business, but not something we are familiar with working in the NHS and universities.

5.6 Non-Executive Director (NED)

There are many opportunities to serve as a NED for a company or an organisation. As a professional, you are likely have specialist knowledge that would be of value to the organisation. Furthermore, your experience in general aspects of management and finance, your high level of analytical skills and experience of working with a range of professionals and stakeholders are likely to make you an ideal candidate for a NED position.

There are many opportunities to serve as a NED in the NHS and these posts are regularly advertised on the dedicated website. If you are interested, I would advise you to make personal contact with the chair of the board of the NHS organisation on which you would like to serve as a NED, and explore the opportunities. In addition to the advertised positions of NEDs, NHS Trusts have the freedom to appoint additional board members who they believe bring valuable specialist expertise.

There are also a number of organisations that you can join, for the payment of a small fee, which will send you NED opportunities that match your specialist expertise and geographic constraints. It is also worth registering with a range of head-hunters who are also often approached by organisations looking for NEDs or trustees.

Should you decide that you wish to serve as a NED, it is important to understand the risks and responsibilities of undertaking such roles. Being a NED can be enormously enjoyable, exposing you to the interesting workings and achievements of an organisation, and giving you the opportunity to work with board members from a range of backgrounds. However, as a NED, you will have some fiduciary responsibilities and you will need to understand how to assure yourself that the organisation is behaving appropriately. A number of organisations offer courses for new NEDs, so it is not too difficult to acquire the necessary knowledge. It is worth noting, however, that being a NED with an NHS Trust is quite different from being a NED for a small company.

5.7   International work

One of the attractions of later-life working is the ability to undertake large-scale international work which may not have been possible when you were holding down a full-time job. As physical and engineering scientists in medicine, you will have specialist expertise which is of great interest in many countries, as well as more general experience in research and innovation, and also teaching and skills development.

Furthermore, during our professional life, many of us would have travelled abroad and had experience of working with a range of cultures, if only for a short period of time at conferences. Some understanding and experience of how different cultures may approach things, plus the respect that many cultures have for older, and by implication, wiser professionals, mean that if you can find a route in, there are opportunities for fascinating international work. In my own case, I have, somewhat unexpectedly, become involved in leadership and management training, specifically in the areas of research, innovation and entrepreneurial activity, in both Kazakhstan and India, which takes me to both those countries for about 6 weeks a year.

5.8   Recommendations

- Decide whether you want to continue to be employed or whether you want to take control of what work you do and the opportunities you take. The latter presents more difficulties but I find it intellectually more challenging and exciting and can recommend it.

- Identify your saleable skills, not your employability. You may think that these are one and the same, but they are different.

- Consider how you will convince people to place a contract with you and how you will assure them that you will deliver what they want on time and on budget. They will be looking at your knowledge base, but more importantly your attitude and behaviour as a professional. Utilising your past connections and contacts is a good start.

- Get into the habit of monitoring how much work you are actually doing for your client. Tracking work to bill a client by the half day is a different mindset that takes some getting used to.

- Explore the opportunities of being an associate for an organisation or company. Increasingly, organisations use paid associates, rather than increasing their own employment base, thus mitigating the risk for an organisation in a changing environment. You will need to identify organisations in the sphere in which you wish to work which have such a policy, and approach them about becoming an associate.

Section 2

Practical issues in research

6 How to get the writing started

Timm Krüger

At some point during your research, it is time to produce a written document. This could be a paper for a certain research community, a literature review for a more general audience, a newspaper article, or a PhD thesis. Writing a long document seems like an impossible task at first glance. How can the writing be started? How can it be structured? In this chapter you will find advice on the writing process itself. Most remarks are so general that you can apply them to nearly all document types.

Many students have limited writing experience. This often leads to fear and procrastination when it comes to producing a scientific document. Adopting a structured approach and considering a few basic rules will help you avoid the writing paralysis and let you produce much better texts.

Note that every person prefers a different reading and writing style. What you see here are suggestions, not a rigid set of rules. You may find some of the paragraphs extremely useful and others less so. You should try different approaches or variations and choose one that suits you best.

6.1   Preparations

The first step is to think about the purpose of your document. Will it be a dissertation or a research paper? Each type of academic document has different 'rules', such as length, structure and level of detail. Longer documents usually take more time and effort to plan and write. If you are working on your PhD thesis, you should allow about 6 months for the write-up. Journal articles can be written in a few days or weeks if the data is already available.

The writing process does not usually begin with the actual write-up. It is crucial to take notes and jot down thoughts during all stages of the research project, including the initial reading phase. Do not wait until the end. Most students have a lab book or send regular progress reports to their supervisors. The lab book should be a repository of all data, ideas and discussions. In many labs, it is mandatory to keep a lab book. Documents such as this are excellent sources of information, even years after writing them. Writing also gets easier with practice. The earlier and the more regularly you write, the more confident you become.

Students may feel overwhelmed by the sheer amount of information they collect during their research projects. The human brain can only hold a few thoughts at any one time. Making notes in a lab book whenever something comes to mind helps to keep the focus on important things. This also includes references and citations.

It is common for students to look for references during the write-up stage. This is a time-consuming and sometimes frustrating process. Questions such as "Where did I read this again?" or "I know it is somewhere in this paper, but on which page?" are distracting and a waste of time. Whenever you encounter interesting data or comments in a paper, a book or the internet, take a note with a short reference.

Although it usually does not matter much whether you keep electronic or hand-written notes, remember that the former are searchable. This is a feature you do not want to miss if you have collected hundreds of pages of notes. Particularly for the management of references it is highly recommended to use tools such as Mendeley or Zotero; these are digital bibliographies that allow you to collect, sort, comment, share and search your collected references. You can also attach the article PDF files and supplementary information. Start using one of these free tools as early as possible to avoid repeating work during a later stage of your reading and writing process.

Maybe the most important thing to realise is that writing a long document is not a single task. Thinking 'I have to write my thesis' does not help to get started. Instead, you should break down the project into manageable pieces. Define subtasks with different time-scales. Writing a chapter might take 2–4 weeks. Each chapter can be broken down into different tasks each of which may take a few hours each. This way you can see the light at the end of the tunnel, and it is easier to begin. The subtasks could be 'define document structure' or 'collect references for Chapter 2'. Sit down with the aim to 'write a draft of the introduction this afternoon', rather than think that you should 'work on the dissertation'.

6.2 Writing environment

Preparation is key to getting into the flow of writing. Still, there are other factors that determine whether you feel motivated. One often neglected aspect is location. You do not have to write in your office or at home. There may be other places where you feel more relaxed and less disturbed. Many people prefer writing in a café or even on a plane. If you are unsure about which writing location suits you well, try different places.

Often it is not the location alone that matters. A mere change of scenery can help. If you have been working in the library for hours, why not take a 15-minute walk to the park, sit on a bench and go through your document structure? Or print a draft of your thesis and sit in the garden to get new ideas. It does not matter where you write; what matters is that you feel comfortable and motivated.

Sometimes you may not be in the mood for working on your dissertation. That is fine. Write when you are feeling energised and use the momentum that builds up. Make sure that you can work without interruption for a few hours. There is no reason to waste your time and generate frustration by trying to write when you are tired or upset. But this is no excuse to procrastinate for days and weeks! If you do not feel motivated for a longer period of time, ask yourself why this is the case. Get back to your document structure and revise your tasks. Maybe they are all too big and appear to be too challenging?

Whenever you feel blocked, take a break. Go outside for a walk or a run and let your mind wander. Get a coffee or a glass of water. Set a time limit to this interruption; 10 to 30 minutes work wonders. Do not take this opportunity to procrastinate, though. Some people avoid breaks because they think they are a waste of time. In reality, not taking breaks strongly reduces your ability to focus. When you are refreshed you will quickly catch up the time you 'lost' during the break.

Sometimes it happens that you lose your focus and motivation for an extended period of time: days or even weeks. This may be a sign of exhaustion. It is crucial that you do not drive yourself into the ground. You are a human and not a machine. There have to be periods of rest after times of hard work and success. If you feel that you are not making significant writing progress for more than 1–2 weeks, take a few days off. Go on holiday if possible. Leave your work on your desk. Having a dedicated time away from your documents during which you can relax is more worthwhile than forcing yourself to stare at your work. You may notice that even a holiday does not improve your motivation. This may be reflective of a level of stress for which you might need to seek help.

6.3   Document structure

There are essentially two approaches to structuring and writing a document: *bottom-up* or *top-down*. You could try both to see which one feels more natural and is more appealing to you.

The **bottom-up approach** means that planning your document and writing its content go hand in hand. In this case, you would initially develop only a rough document structure – introduction, methods, results, conclusions and so on – without thinking too much about subsections. Once you have started your writing (see next section), the structure will automatically evolve. You will realise that certain paragraphs fit better in a different section or that you should split the discussion into several subsections. It is perfectly normal to move sentences, paragraphs or even entire subsections through the document, maybe several times. This is exactly where working on a computer simplifies your life significantly.

If you choose the **top-down approach**, you would think about the structure and the hierarchy of the document before working on the text writing. How many chapters do you think the thesis or report will have? Will the introduction be subdivided into several parts? Which methods do you want to discuss and in which order? How do you want to present your results? Do you want to have separate discussion sections or a single one?

You could then draw a flow diagram of the information you want to convey and play with the structure to make your information as easily accessible as possible. It is helpful to write the table of contents including all chapter and section headings.

You could write a few characterising bullet points in each section. For example, under the section 'numerical methods' you could include the items 'explain why using this algorithm', 'provide short algorithm overview' or 'introduce my own code development achievements'. Ideally, when you have done this for each section, you should have a functional outline of your entire document at hand.

In our experience, the bottom-up approach is more intuitive and flexible than the top-down approach. Writers typically develop new ideas while writing, and a predefined structure can be more restrictive than helpful. Even the most accurately and carefully worked out document structure will probably change over time. However, some people have a very structured working style and may therefore prefer the top-down approach as it provides a clear skeleton to work along.

6.4   The writing process

6.4.1   Hitting the ground running

Many inexperienced writers unconsciously produce text in the order in which it will appear in the final document. They tend to start with the abstract and introduction and then move on to the methods and other later chapters. This approach can easily lead to frustration. You may not be in the mood to write about the methods when you have arrived at the methods section. In fact, there is no reason why you have to force yourself to write in this order.

However, there are good reasons for producing a draft of the introduction early on. Although its final form often only emerges at the end of the entire writing process, sketching the introduction at the beginning forces you to reflect on your actual research aim and objectives. In other words, writing the introduction helps you to see the bigger picture. Without having this deep insight, there is a high risk that your writing will be unstructured and incoherent.

After having completed a first draft of your introduction, you should continue with the section or chapter that is easiest to write. Focus on the section that you feel most comfortable with at any time. Fill gaps later. It may sound counter-intuitive, but this approach maximises your motivation and concentration. However, writing in random order requires you to pay particular attention to the logical flow. The reader should not see in which order you have written the text. This requires you to present the final text in a coherent fashion. Do not worry about that during the initial writing phase, though. During the later editing stages (see below), you will identify and remove the artefacts caused by the random writing order.

When you work on your document, it is tempting to get each sentence right before you move on to the next. This approach usually disrupts the flow of writing if you cannot find the appropriate word or expression. Read a single paragraph of any paper you have on your desk. Do you really believe that the authors wrote all paragraphs in a single go? The truth is that you do not know how often the authors revised their paragraphs. Reading published articles creates the impression that writing is a single-pass process. It is not. Most scientists need several iterations of writing, editing and polishing before they have produced a text that is appealing and readable.

One of the key pieces of writing advice is to keep 'writing' and 'editing' separate. When you write, focus on the content, not the words, typos or logical gaps. These can be fixed later. Remind yourself from time to time that you do not have to produce the final text in your first pass. In the first instance, make sure that you get the information on the paper (or on the screen). You can even leave out entire sentences or arguments and keep notes such as 'to do', 'fill gap' or 'look up details in the literature'. Once you have written anything down, you can come back later to revise and edit it. This writing approach may not seem perfectly satisfactory at first, but it keeps you motivated and your mind uncluttered.

If you cannot make a start with your writing at all, try to write what you are thinking about at that moment for five minutes without stopping. Your text does not have to make any sense. You can delete it later. The purpose of this exercise is to put your brain in writing mode. Like in many other situations, the real hurdle is to get something started. Once the ball is rolling, it is easier to keep it moving. This is also the reason why you should work on simple paragraphs if you are not in the mood for writing more challenging pieces of your document. Once you have finished an easier section, you may feel confident enough to attack the more challenging ones.

6.4.2 Writing for the readers

During your PhD project or research activity, you will have performed many experiments or simulations. You have probably developed a new code and performed many calculations. You have written down countless ideas, some of which may actually lead to scientific breakthroughs. But to be honest, not everything you produced should end up in your dissertation or report. Maybe you performed a series of experiments with wrong settings. Maybe you did a long calculation, just to get a trivial result in the end. Who would be interested in reading these paragraphs? What would they contribute to the scientific community?

Not all information should go into your final document. This is probably one of the most painful aspects of the writing process, yet omission is absolutely necessary. If you feel bad about removing calculations or data from your notes, you should keep them in a second document, a 'not-to-publish' or 'dump' text. For example, if you have written a rather technical summary of a specific aspect of your work that would only distract the reader, move it to the dump document. You may recycle it for a more specific paper in the future, or you could use it to explain your method to colleagues in an email. This is all about psychology: we do not like to delete things we have written because it took a great deal of time and effort to create them in the first place. Knowing that your text will not be deleted for good helps you keep your document clean and concise.

It may come as a big surprise to some people that you are not writing your scientific texts for yourself. Your texts are meant to be read and understood by others, not your clones. Depending on your document type, your readers may be your peers, your supervisors or funders. A remarkable number of scientists do not understand this rather obvious point. The real challenge of scientific writing is to convey your messages in such a way that your readers do not get lost. Consider what the reader may think when he or she reads your sentences and paragraphs. Think about newcomers and people from other fields. Your writing style has to reflect that there is an increasingly interdisciplinary research landscape.

You are a reader yourself. What do you expect when you read a paper? Why do some papers annoy you when you read them? There could be a lack of clarity, no motivation provided or illogical connections between sections. As a rule of thumb, avoiding what you do not like about other authors' works and imitating what you like is a good way to improve the quality of your own text.

Before you started your writing, you thought about the structure of your document. When you write, stick to your outline. Do not digress. If you have decided not to write about 'experiment 4B', why would you suddenly change your mind and write about it? Only because it seems to be easier to explain than other things? If you realise that you are drifting, force yourself to go back to the topic. You can only decrease the quality of your text (and make the readers impatient) if you include unimportant details. Changing your document outline during the write-up is sometimes necessary, but always ask yourself whether those changes increase the quality of your work.

Use figures and diagrams to highlight key findings and data. A single figure can replace a thousand words. If you have used an experimental device, explain its principles of operation in illustrations. If you simulate blood flow in a complex geometry, show the geometry. Create simple sketches of the figures early on, and keep the polishing for the editing stage. If you prepare a shiny and detailed figure at the beginning, your understanding or interpretation of the data may still change during the write-up and you may have to change or redo the figure. This can be an enormous waste of time. Avoid creating unnecessary figures or showing every single data set. Taking your figures and tables alone, they should tell a short version of your entire story, but not more than that.

Put information where the reader expects it. Have a logical structure throughout. Numerical methods are discussed in the methods chapter, not in the results sections. Important definitions and concepts that you use throughout your text, such as 'magnetic resonance tomography' or 'wall shear stress', should be discussed in the introduction or background sections. Readers become easily annoyed and impatient when there are random jumps in the flow. If there is some rather technical detail you would like to include without interrupting the main text, put it in the appendix and refer to it whenever required. If you write a dissertation or another long document, an index helps you and the reader to find relevant information easily.

Do not assume that the reader will understand implications. Be explicit. If you have an ambiguous statement in your text, 20 different readers will interpret it in 20 different ways (and probably not in the way you intended it to be understood). All readers actually see your text differently; it is your job to minimise the risk of having vastly divergent interpretations. For example, instead of writing 'the Navier-Stokes equations have been solved numerically', explain which method has been used, refer to other sections with more details or provide literature references.

6.4.3   Getting early feedback

Give your text or single sections to other people. It is useful to get feedback at any stage of your work. For example, you could show your supervisor the document structure you have come up with. Your document has a certain target audience. Members of this target audience are the most valuable reviewers, in particular, if they are experienced in the topic or writing in general. If you write a journal article, say, give a draft to one of your peers. Be aware that many people prefer advanced text drafts. Talk to your supervisor and peers, and ask them whether they are happy to review rough text drafts as well. Also bear in mind that many people will not read your text over and over again, so think about what the best time is to send your draft to them.

When you have finished writing your first draft, put it aside for a few days. Having a fresh mind and some distance from your text is beneficial for the editing process. Relax and treat yourself. You have earned a break.

6.5   The first editing stage: clarity and conciseness

After writing your first draft – this can be your entire document or just a chapter – you can edit it. Many people edit their texts in multiple stages. For example, during your first revision you should try to make sense of the text. Are all arguments clearly explained? Is the order of your sentences meaningful? Do you show all necessary data? In a later pass, you can simplify your text as much as possible. Are all sentences relevant? Have you used complicated words that could be replaced by simpler ones?

Are there repetitions? Finally, you can focus on the finer details, such as punctuation, typos and style (see below).

When you edit your text, pretend to be a reader of the target audience. If you worked on an experimental technique for 3 years, you will hopefully understand it better than most other people. Your readers will probably not. You cannot assume that they have the same level of knowledge and understanding. Would somebody who has not worked on this problem for 3 years understand what this particular sentence or paragraph means? If there is a logical connection between two points, write it down. Do not assume that the reader will understand your points somehow. If there is something the reader should know, tell him or her.

You can practise this writing style by reading other people's texts more consciously. What do you admire about the style of other papers or dissertations? What don't you like? Whenever you find something remarkable (either positive or negative), take a note and try to incorporate or avoid this style in your own writing.

Avoid repeating definitions and explanations. You can (and should) always refer back to relevant sections instead. In some cases, you can also refer to a later section, e.g. to provide an outlook. If A is needed to understand B, then you should write about A before B. If you talk about 'tissue elasticity' several times, define it in a background section and refer to it when you need to. Add an index item if appropriate. If the reader already knows what is going on, he or she does not have to go back and the text is not interrupted by a repetition. If the reader wants to know more, following your well-placed cross-references does the trick.

Each sentence has to be clear. If it does not have a meaning, delete it. You do not get extra marks or recognition for additional sentences. If a sentence conveys two things at the same time, split it into two sentences. Make sure that your sentences are logically connected. Each sentence should naturally lead to the next one. If this connection is missing, the reader can easily get confused.

Keep it short. Make clear what the main points of your reasoning are. The reader prefers a concise document rather than a bloated one. You will not improve the quality of your research by writing more about it than necessary. Be careful, though! Writing more concisely is more difficult, but it is worth the effort. If a concept is difficult to explain, writing more about it seems to be a good solution. In reality, it is more helpful to think carefully about the nature of the concept and find better words to describe it. Talking to your colleagues and friends usually helps to find a better way of explaining things. Their reaction tells you whether you have used the right words or not. Sometimes you need to look at your text with additional distance. You could continue editing another section first and come back to the problematic bits later on.

Be critical about the length of your chapters and sections. The length depends on the relevance of each part. If you are using well-known methods to achieve new results, your results section will typically be longer than your methods section. If you are primarily developing new algorithms, you could have a large methods chapter. An experienced reader will spot bloated sections (and be disappointed or annoyed by them). Compare the relative importance of your chapters with their actual length. Does the length of each section appear reasonable?

Keep it simple. There is no need to use a complicated explanation when a simple one does the trick. You are not rewarded for writing complicated text, although some authors seem to think so. Readers will appreciate the accessibility of your text more than your ability to write Latin or technobabble. Use simple words. Your scientific achievements should speak for themselves. They do not benefit from over-complication.

6.6   The second editing stage: language and style

There are countless books and articles about writing style, language and paragraph structures, such as 'The Elements of Style' by Strunk and White, or 'On Writing Well' by Zinsser. Here we only collect a few important points. Although this list is far from being complete, sticking to it helps to significantly improve the quality of your text.

- Use active sentences whenever possible. Scientific texts make heavy use of passive constructions because the object, not the scientist, is the focus of the research. Yet, you can change passive to active in many cases. For example, replace 'An improved method was proposed by Doe' by 'Doe proposed an improved method'.

- There is an ongoing discussion whether to write 'I' or 'we' in scientific single-author texts. Some people prefer avoiding both pronouns altogether. For example, instead of writing 'I can conclude' you could take 'One can conclude' or 'It can be concluded'. While there is no clear rule or generally accepted recommendation, you have to be consistent. If you use 'I' in one chapter, stick with it everywhere else.

- Use verbs to replace nouns. Write 'The improved code minimises numerical instability' instead of 'The improved code leads to minimisation of the numerical instability'.

- Avoid unspecific quantifiers such as 'very' or 'few', in particular when you have hard data. What does 'very expensive' or 'very accurate' mean? Those words cause speculation and interpretation and reduce the credibility of your work.

- Each paragraph should begin with a sentence that lets the reader see what the paragraph is about. Discuss only one idea or concept in each paragraph. If a paragraph is too long, the reader tends to scan forward; 5–10 lines is a reasonable length for a paragraph. Each paragraph should lead to the next in a logical way.

- Keep the tense consistent, especially in descriptions and the methodology part. Although there is no fixed rule, past tense in the methodology and present tense in the discussion are reasonable choices.

- Read your written text aloud. If there is a sentence that sounds strange, rewrite it. The quality of the written text is usually higher when it sounds comfortable.

- Use a spell checker!

- Finally, put the document away for a few days. Read everything again and give it a final polishing round.

As mentioned at the beginning, this chapter is based on the personal experience of the author. No size fits all, so you may disregard some of the suggestions presented above. Your writing performance is best when you feel comfortable. In the end you have to find out, through continued practice, which writing rituals and habits give you the most satisfaction. But in any case: the more you write, the easier it becomes.

7 The PhD thesis

Peter R. Hoskins

The PhD is a research degree which culminates in the production of a thesis written by the student which is then examined. Here we discuss the PhD thesis and, in the following chapter, the oral exam (the viva) and its aftermath.

7.1 PhD and thesis aim

For those who have read previous chapters in this book, it should come as no surprise that the PhD thesis needs an overarching aim, that the PhD thesis should be constructed to address this aim, and that there should be a conclusion at the end of the thesis which relates to the aim. When the student starts a PhD, there is usually some general area of research with 1–2 research questions which have been produced by the supervisors, usually summarised in a single page of A4. PhDs should, at least to some extent, be open ended allowing the student to follow interesting avenues as they appear. Some things which looked promising at the beginning of the thesis will turn out to be impossible, uninteresting or too difficult. No matter what changes in research direction there have been during the PhD, the thesis needs a clearly defined overarching aim which can be expressed in a single sentence; 'The aim of the PhD is ...'.

Following on from this, the contents of the thesis should form a coherent whole related to the aim. Once the student starts to write their thesis, usually sometime in the final year, it may become clear that the overall direction has shifted and that some of the work undertaken may now be difficult to fit within the overarching theme of the thesis. This is unavoidable and the exclusion of such work does not constitute bad planning or failure. This is an inevitable consequence of research which by definition is concerned with doing things that have never been done before and finding out things which were previously unknown. In this landscape, it is impossible to define exactly where one will end up.

With the above in mind, the PhD student needs to have an idea fairly early in the PhD on what the overarching aim of their PhD is, and be prepared to make changes to the overarching aim as the research progresses. A clear sign of problems is if there is silence when the PhD student is asked what they think their overarching aim is. PhD examiners may ask this question and expect an immediate short answer.

7.2   Thesis structure

The final written thesis in medical physics or bioengineering is typically 200 pages of A4 (single side) with around 40 000 words (excluding front-piece material such as title page, contents, declaration, etc.). Universities impose an upper limit somewhere in the range 80–100 000 words. Theses whose length is close to the maximum word count may require two volumes. This is not advised. Imagine the examiner of a PhD thesis receiving two volumes in the post and inwardly groaning at the length of time that it will take to read this.

PhD theses from one department may be structured in a similar manner. Supervisors end up sharing common practice and students will look at previous PhD theses from the same department and copy the style. This can lead over time to theses from the same department sharing a common structure. The thesis structure below corresponds to that from Medical Physics in Edinburgh. Other structures will be briefly commented on later.

- Introduction; 20–40 pages.

- 4–6 chapters of original work; 120–160 pages

- 1 chapter of discussion conclusion and future work; 5 pages

- Appendices as necessary (e.g. software, questionnaires); 5–10 pages

- References; 15–25 pages

- List of papers and conference presentations arising from the thesis; 1–2 pages

The overall length usually lies between 180 and 230 pages. The original work should obviously form a substantial part of the thesis; around two-thirds of the whole. This is a simple metric and obviously no guide to quality. However, if the examiners are presented with a thesis where the original work is less than 50% of the whole, then this raises immediate questions about whether there is sufficient material to justify awarding a PhD. Where the examiners confirm that there is insufficient work, then the viva result may be to resubmit when more work has been done and included in the thesis.

7.3   What to put in the thesis

Suggested details of what should be included in the various sections of the thesis are discussed below.

Introduction

The purpose of the introduction is to define the research area in sufficient detail to justify what the aim of the thesis is. The aim of the thesis comes near the end of the chapter and should be written as a separate paragraph in the form: 'The aim of the thesis is...'. Some students may put this in bold so it can be easily found by the examiner.

The introduction should be written in a top-down manner, starting broad and gradually narrowing down to the specific area of interest. For example, a thesis on the use of a new Doppler ultrasound technique for diagnosis of carotid artery disease might be structured as follows:

1. Section on demographics of world deaths stating that cardiovascular disease is responsible for one-third of all world deaths and quoting WHO references.

2. Section on carotid disease; prevalence, pathophysiology, diagnosis and treatment.

3. Section looking at current methods of diagnosis using ultrasound with an emphasis on Doppler ultrasound techniques.

4. Section on recent research work on Doppler ultrasound techniques.

5. Detailed section with sufficient maths to understand issues of specific techniques relevant to the thesis.

6. Critical analysis of the state-of-the-art research in Doppler ultrasound, leading to ...

7. Aim of thesis written in the form: 'The aim of this thesis is ...'.

8. Section beginning: 'The thesis is structured in the following manner to address this aim:'. This is followed by brief summaries of the subsequent chapters, e.g. 'Chapter 2 is concerned with ...'.

In the above list, sections 1–6 should be referenced with review articles and key papers found from the literature. The literature review aspect of chapter 1 is not just to describe the literature; it is to point out where gaps are present in the literature and where there is a need for new work. This can be undertaken, for example, at the end of each section. It is however useful to have a specific section (listed as 6 above) in which the student critically analyses the state-of-the-art. This then leads logically to the aim of the thesis which follows immediately after section 6 in the list above.

A common mistake is to include too much basic bookwork. The thesis, and especially the introduction, is not a textbook. Background technical details can be included but there should be some reason for this. The student should ask themselves how does including technical detail help define the aim of the thesis? In the above example, it would be relevant to include details on the basic physical principles by which blood velocity is estimated using Doppler ultrasound. It would not however be relevant to include large amounts of technical details which were not relevant to the thesis aim, such as beam-forming, transducer design, etc. Inclusion of large amounts of textbook work usually results in requests to remove most of this and make the remaining text more relevant to the thesis.

Chapters of original work

The PhD usually consists of severable identifiable pieces of work, each of which can form a paper or chapter. For a PhD in which some new technology is developed, the various phases of work might involve the following work-packages:

- development of the new technology (hardware, software),

- development of experimental (laboratory) systems to validate the new technology; often called 'phantoms' in medical imaging or modelling,

- experiments to validate the technology,

- applications in patients; either initial studies to demonstrate that the technology works or longer studies to answer a clinical question.

Each of these work-packages could form a separate chapter. Each chapter may be written up in the manner one would write a paper: introduction including literature review relevant to the chapter, aim of chapter, methods, results, discussion and conclusion.

It is worth further exploring the difference between the literature review in chapter 1, and the literature review in each of the chapters of original work. In chapter 1, sufficient literature is presented in order to justify the aims of the thesis. In the chapters of original work, the literature review is mainly presented in order to help justify the choice of methodology.

Chapter on discussion, conclusion and future work

This chapter consists of three sections separately labelled: discussion, conclusion and future work. Each of the individual chapters of original work should have a discussion relevant to the chapter. In the 'discussion, conclusion and future work' chapter, the discussion of the original work should be with respect to the thesis as a whole. Note that a discussion is not a summary of the main findings. The discussion should include the following issues: methodological concerns which may affect results and interpretation, difficulties encountered during the thesis which impacted on the work, limitations of the work, and comparison with the work of other groups. It is best to stick to the main points and keep this section short, 2–3 pages maximum.

The conclusion is not a summary or abstract and is not the place to repeat the main findings from every other chapter in the thesis. The conclusion should specifically be related to the aim of the thesis and be a few lines (1/4–1/2 page) as to how, overall, the thesis addressed the aims set out in chapter 1.

The section on future work is an opportunity to explore what might next be done to take this area forward, by say another PhD student.

Appendices

Appendices can be included in the thesis though these are not essential. In some departments, it is practice to include printouts of code for key programs written by the student including sufficient commentary to be understood by the examiner. This provides some form of evidence that it is the student that has written the code. Other documents can be included such as questionnaires. Some appendices may include methodology which the student has deemed to be too detailed to include in the chapters of original work. The author's opinion is that these should be included at the end of the relevant chapter (or not at all). Some theses include very large appendices of raw data, however, this is bad practice. There is no need for this; each chapter should provide sufficient detail of data in the results section without the need for more material in an appendix.

References

This is a list of all papers, patents and other documents referenced in the thesis in alphabetical order. The host university will provide guidance on the referencing format.

List of papers and presentations

All papers published or in-press should be mentioned along with a list of oral and poster presentations arising from the work of the thesis.

Additional material

In addition to the above, a CD or DVD may also be included in the thesis tucked into a pocket on the inside or back cover. This could include examples of key software and multimedia elements such as videos and animations. The student may also choose to put material on a website.

7.4   Alternative thesis structures

The above thesis structure is based on successful PhDs from Medical Physics in Edinburgh. Other structures which the author has seen in examined PhDs are listed below.

7.4.1   Chapter 1 as an extended abstract

This is a very short chapter of 3–4 pages at the beginning of the thesis which could be considered to be an extended abstract. The purpose of this chapter is to provide a brief overview for the examiner of the main background (often with no or minimal references), the aim of the thesis, the main structure of the thesis, and in some cases, the main findings of the thesis. The rationale is that the examiner knows where they are going and knows what to expect in the remainder of the thesis.

The aim of the thesis should come from presentation and critical analysis of the current state-of-the-art. In this example, the aim appears from almost nowhere and can lead the examiner to make comments that the aim is unjustified and that the background literature is insufficient. If the full literature review appears in the second chapter, this can look confusing; why has the aim been defined in chapter 1 when the literature review is in chapter 2? It is also more difficult to connect the literature review with the aim unless the aim is repeated at the end of chapter 2. This may also look odd to the examiner who wonders why there is an aim in chapter 1 and a repeat of the aim in chapter 2.

The author advises that if a short chapter 1 is included, then it should be made clear that this chapter is an extended abstract meant to orient the examiner and that a full literature review justifying the aim is presented in subsequent chapters.

7.4.2 Entire thesis written as a big paper

In this structure, all methods are in 1–2 chapters, followed by all results in 1–2 chapters followed by a chapter of all discussions and so on. If the thesis has been concerned with one big work-package leading to one result, this structure might work. Most theses contain several packages of work so that a thesis structured as one big paper is almost impossible to read from an examiner's point of view.

7.4.3 The textbook

This is where the first few chapters are written in the form of a textbook, usually with no explanation of which sections are relevant to the aims of the thesis. This is not recommended. However, it is worth trying to understand why students structure their thesis in this manner. One possibility is that the student does not understand what undertaking a PhD is about, or what should be presented in the thesis. If a PhD is around 200 pages, then writing 100 pages of textbook material is half the PhD thesis – right? (Definitely not right.) The other possibility is that the student (and possibly supervisors) are aware that there is insufficient original material and are simply using a long introduction as a form of padding. Examiners will not be fooled by this and textbook PhDs are very likely to be sent back by the examiner with the requirement for considerable rewriting.

7.4.4 Inclusion of systematic review chapters

Some PhDs include an element of systematic review. Systematic review is a form of literature review defined as 'a review of the evidence on a clearly formulated question that uses systematic and explicit methods to identify, select and critically appraise relevant primary research, and to extract and analyse data from the studies that are included in the review'. This is classed as secondary research (making use of existing knowledge) rather than primary research (generating new knowledge). This type of literature review would appear as a separate chapter of 'original work' rather than in chapter 1.

7.4.5   Thesis incorporating publications

Some universities allow published journal papers to be included in the thesis in a more direct way. The thesis includes text which contextualises the paper in terms of the paper's place in the thesis. Generally only papers which have been accepted for publication can be incorporated in this manner.

7.5   Signposting

The student will probably know every section of their thesis from beginning to end and if asked where in the thesis a particular method or data is will be able to turn to that page. Unfortunately, the examiner will not have the same knowledge of the thesis. It is best to imagine the examiner as a busy person who is trying to read the thesis in what little time they have. It should be clear to the examiner why they are reading a particular section. If after a few pages, the examiner thinks to themselves 'why I am reading this' or 'I don't understand what the student is trying to tell me', or even 'I don't believe any of this', then this is not good. This will lead the examiner to go backwards and forwards in the thesis looking for key explanations. This is time consuming, frustrating and annoying and an annoyed examiner may be more likely to view the work negatively.

Signposting consists of putting in sentences throughout the thesis so that the examiner does not get lost. At the beginning of each chapter, it is a good idea to put in a sentence beginning: 'This chapter will describe ...'. Each major section can have a similar sentence: 'In this section will be described ...'. At the end of the section, a paragraph could be included which provides a brief summary of the key points from the section. In this way, at the beginning of each new chapter/section, the examiner knows what to expect and will be looking for this. At the end of each section/chapter, the examiner is told the key points that the student is trying to get across. In other words, the examiner is told where to go at the beginning and then told where they have been at the end.

7.6   When to start writing the thesis

A 3-year PhD can broadly be divided into 6 months getting up to speed and pilot studies, 2 years data collection and analysis, and 6 months writing the thesis. Having said this, there will not be a clean divide between finishing the practical work and beginning writing the thesis. The more likely scenario is that the two will run in parallel, usually almost up to the time when the thesis is handed in. The student should start writing the thesis sometime in the final year, and certainly be writing at least 6 months before the expected hand in deadline (usually when the funding runs out).

Starting a 40 000 word thesis can be a daunting prospect. Just to give this some perspective, a typical academic might generate 1–2 first author papers or chapters in a typical year which is around 5000–15 000 words. A PhD represents an extremely large writing task which most academics will not repeat unless they write a book (50 000 words a year over 2 years for a 100 000 word book). The student should expect to feel writing-overload towards the end. For this reason, it is best if some of the writing can be done earlier in the form of reports or papers. Writing papers as the research progresses is an especially good thing to do for the student. Each paper is a readymade chapter; the paper content can be cut-and-paste into the chapter with relatively little modification. Writing papers is also a sign of quality in that the work has been externally reviewed. It is very hard for an examiner to trash or reject work in the thesis which has already been published as journal papers.

Students where English is not their first language have extra challenges associated with writing in a foreign language. It is advisable that these students attend courses on writing early in their PhD and also start writing reports on progress from an early stage. The thesis should be the student's own work and most supervisors are unwilling to spend large amounts of time changing text to make this grammatically correct from an English point of view.

Some students from an Engineering and Physics background may not have been exposed to essay writing since they did English or History at school aged 16. Where engineering and physics undergraduate degree courses have concentrated on problem solving, then the student may be unfamiliar with the discipline of essay writing. The situation is most challenging when the student has both lack of essay writing experience and also does not have English as their first language. In all cases, early exposure to report writing in the PhD is essential.

7.7   Summary

The PhD thesis represents the final output of the PhD and is examined in the viva. The PhD is best structured in a top-down manner starting with a chapter which reviews the literature and defines a research aim, chapters of original work designed to address the aim and a conclusion which relates to the aim. Writing the thesis is time consuming and the most efficient way is to write papers following the completion of each work-package, so that papers can be cut-and-pasted into chapters. Those students who have not had recent experience of essay writing and/or do not have English as a first language are advised to start report writing early in the PhD and to undertake training in scientific writing in English, which is offered by most research-intensive universities for precisely this reason.

8   The PhD viva

Peter R. Hoskins

The PhD student has finally, after huge effort, handed in their PhD thesis to the university and now waits for the exam. The exam is called a 'viva-voce' which translates as 'with living voice' which means that this is an oral exam (referred to henceforth as the 'viva') rather than a written exam. This chapter outlines what usually happens in a viva, how to prepare and behave on the day, and what happens after the viva.

8.1   Arrangements for the viva

The main parties in a PhD viva are the external examiner (i.e. external to the host university), the internal examiner and the student. Both internal and external examiner have a role in the examination. The final outcome is agreed jointly between both internal and external. Traditionally, there is a hierarchy in that the external examiner is seen as the main examiner and the internal as a secondary examiner who chairs the viva, ensures that the exam is conducted according to the university guidelines and makes sure that the paperwork is submitted to the university. These traditional roles have shifted slightly with the growing number of interdisciplinary PhDs (especially in bioengineering) where the student may have developed technology in two or three areas, and in some cases, undertaken both engineering/physics and biology developments. Identifying an external examiner to cover all the relevant technology areas may be challenging. However, if the external examiner covers the main technology areas, the remaining areas can be covered by the internal examiner. For example, for a PhD concerned with integration of MRI with computational modelling, the external might be someone involved in patient specific modelling who is an expert in modelling, and the internal might be an MRI physicist. One examiner may also come from a medical or biological background to examine on the application aspects of the PhD. For interdisciplinary PhDs, the internal examiner plays a far more important examining role than is the case in a single-technology PhD. The above model of a single internal examiner and a single external examiner is the most common. However, there are other models. Some universities do not have an internal and instead, there are two external examiners. Where the PhD student is a staff member, some universities require two external examiners and an internal examiner. In some universities, there is a non-examining chair. This may be for all PhD vivas or just for those PhD vivas where the internal or external examiner are relatively new to examining.

Supervisors are allowed to attend the viva if the student wants this. The student has the right to say no to requests from the supervisors to attend. The supervisors are usually not allowed to speak during the viva; in some universities, they must remain silent throughout the exam, in others they can speak if invited to do so by the chair or by the examiners. Some supervisors can end up answering the examiners' questions on behalf of the student or trying to defend the student. This is unacceptable and the chairperson should stop this. The need for the chair to intervene in this manner will clearly raise the level of stress in the room. In the experience of the author, supervisors are best excluded from the viva. If the student does feel the need for some kind of moral support, then only one of the supervisors should be asked, and should sit and say nothing at all.

The formal process for selection of examiners begins when the student is 3–6 months from handing in. The student fills in an 'intention to submit' form and submits to the university. In some universities, the examiners are named on the intention to submit form; in others, a request for names is sent to the principal supervisor or Head of Department. While it is possible for the Head of Department or principal supervisor to pick examiners without consulting the PhD student, in most cases, this is done after discussion amongst all supervisors and with the student. Generally, the examiners are approached informally by the principal supervisor so that they are expecting the formal invitation from the university. Usually there is an approval process to make sure that the proposed examiners meet the requirements of the university regulations and are academically suitable.

Once the student has handed in their PhD thesis (2–3 copies) to the university, then the supervisors no longer have responsibility for the thesis. The university communicates with one of the exam team from the host university (e.g. the internal examiner or the non-examining chair) concerning the viva, who in turn communicates with the external and the student concerning the viva. Usually the viva occurs 2–3 months after handing in the thesis. A delay of 6 months would be considered to be long. PhD students from abroad often prefer to have a very early viva, while their visa is still valid, in order to avoid the cost of returning to the host university from their own country. Examiners usually try to accommodate this but this can be challenging and the examiners have the right to say no to an early exam.

8.2   How to prepare for the viva

There is a delay of around 2–3 months between handing in the thesis and the viva. Most students will be fairly tired at the point of handing in and most will want to take a well-deserved holiday. There is no necessity to spend all of the time between hand-in and viva going over the thesis. The author's recommendation to his own students is that, in the 1–2 weeks before the viva, it is worth reading the thesis a couple of times for familiarisation. If the student realises that there are areas where they are not quite clear, then the student should spend some time re-reading the key papers just to try to understand these. However, there is no need to do large amounts of extra work. The main work has been done in the process of writing the thesis; there is no need to repeat this.

If the student has time, the best thing that they can do is to continue to write and submit journal papers. This will help the student's CV and will help the supervisor's CV. It will also help in the viva where the student can give an update on the current state of journal papers, even bringing in emails from the journal with confirmation of paper submission or acceptance.

Some centres undertake a mock-viva. This is useful preparation and can be undertaken at some point during the final year while the student is still present in the department.

8.3   The viva

The viva consists of a series of questions from the examiners which the student must answer. Some universities require a presentation to an audience of the work of the thesis before the viva. Other universities have no audience participation, but the examiners may ask for a presentation describing the key aspects of the thesis. Others have no presentation and just consist of questions to the student. The length of the viva is variable, usually between 2 and 4 hours. A viva of less than an hour does occur as does a viva of over 6 hours. A very short viva suggests that the student has not been properly tested. A very long viva may occur in a borderline PhD where time needs to be taken to go through the work. The author has heard of a very long viva where the student was very nervous and when exposed to an aggressive examiner was unable to articulate an answer adequately. When it is clear that the viva is going to take a long time, the chair or internal examiner usually makes arrangements for meal breaks.

Most students are slightly apprehensive at the start of the viva, some extremely nervous. Most examiners will start with a series of general questions designed to get the student talking and to begin to relax the student. Giving a presentation helps with this process. However, there are variations in this and some examiners have no interest in the emotional state of the student, seeing their job as to put the student on the spot in order that they can defend themselves. The PhD viva is, after all, also called a 'thesis defence'.

Following general questions, most vivas take the student through the thesis starting at the beginning of the thesis and going to the end. Usually, examiners have a pre-prepared list of questions and/or a copy of the thesis stuffed with stick-it notes in particular pages.

Most students will relax after a while and the viva often becomes a dialogue between scientists. After 3–4 years of intense study in an area, the student should know the subject of their thesis very well and be able to have a high level scientific discussion with experts.

If the student does not understand the question, it is best to say 'I don't understand this, could you ask this in a different way'. If the student does not know the answer, it is best to say 'I don't know' rather than waffle.

There are a few circumstances in which the viva chair needs to step in and either stop the exam or call a break, though these are uncommon occurrences. Some universities have a rule which is that if the student states during the exam that failure of the viva will have negative consequences for the student (e.g. paying back loans) then the viva chair must stop the exam immediately.

When the examiners have finished questioning the student, the student is asked to leave the room while the examiners consider the outcome. Usually the student is brought back in and told the outcome (covered in the next section). The viva chair or internal coordinates a reply to the university who then conveys this to the student.

8.4   After the viva

The viva outcomes are university-dependent and the list below is typical:

i) Award PhD. No changes needed.

ii) Minor corrections needed. Corrections of an editorial nature are needed, but not more work. The student is given 3 months to complete, the internal examiner signs off the revised thesis and the degree is awarded.

iii) Major corrections needed. Corrections are needed some of which include more work. The student is given a time between 3 and 6 months to complete, the internal examiner signs off the revised thesis and the degree is awarded.

iv) Major corrections and re-examination. Major corrections are needed which require substantial reworking of the thesis. The student is given between 6 and 12 months to complete, and resubmits a revised thesis for a second viva.

v) Award an MPhil. The thesis is substantially deficient and in the examiners' opinion could not be revised to reach PhD level.

vi) Award MPhil following minor correction.

vii) Major corrections and then re-examine for MPhil.

viii) Award MSc by research.

ix) Fail.

The most common outcomes are ii) and iii): minor or major corrections to the thesis reviewed by the internal after 3–6 months and the PhD awarded. Other outcomes become less and less common as one progresses down the list. In the case of 'Fail' in the UK, this happens around once in 400 exams.

Not all students who start their PhD will hand in a PhD thesis. In the review towards the end of the first year, the decision is made to allow the student to continue. This is the principal point at which PhD students discontinue or continue with a lower degree. If the student has made good progress to this point, it is likely that they will continue on to hand in a thesis. However, things can change between the end of first year and the end of the PhD. It may be that the supervisors judge that the PhD thesis is not of sufficient quality to justify the award of PhD. Any lack of PhD progress should be flagged at second and third year reviews and in supervision meetings, and attempts should be made to get the PhD back on track. However, in those cases where the supervisors judge that a PhD is unlikely to be awarded, they should advise the student of this and recommend submitting a thesis for a lower degree. Some cases of PhD fail are where the student has been so advised but insists on carrying on to the PhD viva.

Usually, the student's funding period will have finished by the time of the viva. Very few students hand in sufficiently early to have their viva while still being funded by the PhD studentship. In this situation, students will have moved on to the next job and/or returned to their country of origin. If the outcome of the PhD viva is to undertake some rewriting, then the student can usually undertake this from their new location without too much problem. The student retains access to university library facilities and will either have access to their old computer via remote login, or will have taken all of their files with them to use on their own computer in their new location.

Any outcome requiring the student to undertake more lab work may be problematic for the student. The funding has run out so the student either has to self-support during the extra months of work, or (rarely) the supervisor provides some funding from their own work funds. The necessity to undertake more work is especially problematic where the student has moved on to a new job/country. If the work is computer-based (e.g. computational modelling), then the student may be able to undertake the work remotely, provided they can continue to have login access and that software licences remain up to date. If the work is lab-based, then the student may need to make arrangements to visit the university for 1–2 weeks including travelling to and from their new location at their own expense.

Once the thesis has been signed off by the examiners, the student hands in the final thesis to the university, usually both hardcopy and electronic copy.

8.5   Summary

The PhD viva is the culmination of the PhD and is an experience most students remember for the rest of their lives. The most common outcome of the viva is the award of the PhD after minor editorial changes to the thesis, or (less commonly) award of the PhD after completion of more work. Other outcomes such as the award of a lower degree or even fail are relatively uncommon.

9 Paper-driven research

Peter R. Hoskins

Previous chapters have described the academic career discussing the expectations on the individual at each stage of their career. There has been some mention of successful strategies for going about research, but this has not been covered in any detail. This chapter concentrates on the detail of how to do research, highlighting strategies which (in the opinion of the author) are successful and those which are not. Paper output and research should be done in unison, rather than the paper output being an afterthought. This chapter will explore the link between research and paper output and discuss strategies which best maximise paper output.

9.1   Is it research?

Let us start with the activities listed below. Ask yourself if these activities are sufficient to constitute research.

- paper reading
- data collection
- developing hardware and software
- seeing your supervisor
- attending lectures
- giving project update meetings
- getting frustrated
- chatting to colleagues
- moaning about supervisors/researchers/University/EPSRC, etc.

These are typical activities undertaken by researchers in medical physics and bioengineering. To further consider if this list is sufficient to constitute research, let us look at different definitions of research from the web:

- 'A detailed study of a subject, especially in order to discover new information or reach a new understanding'. Cambridge Dictionaries Online.
- 'The systematic investigation into and study of materials and sources in order to establish new facts and reach new conclusions'. Oxford Dictionaries.

- 'Performance of a methodical study in order to prove a hypothesis or answer a specific question'. Explorable.com.

- 'The attempt to derive generalizable new knowledge including studies that aim to generate hypotheses as well as studies that aim to test them'. NHS Health Research Authority.

These definitions highlight the importance in research of defining an aim (hypothesis or a research question), undertaking a systematic study designed to address the aim, and coming to a conclusion related to the aim. We can now look at the list of activities above and conclude that this list is not sufficient to constitute research. It is research when:

- there is an aim,

- there is a conclusion.

9.2   Defining the aim

Defining an aim is central to research. One might ask where the aim comes from. Those in the best position to define a research aim which is worth pursuing are obviously those with some experience in the field who know what the state-of-the-art is. The academic should know what the current state-of-the-art is in their field which should enable them to come up with new research directions and associated research aims. Translating this into grant funding is a different challenge not addressed in this chapter! A decent research question therefore comes out of knowledge of the subject which, in turn, comes from knowledge of current work in the area by reading the research literature and by attendance at conferences. PhD students are usually presented with a research area in which there are a few broad aims. The student has an opportunity to refine and extend these aims depending on their own interests and on how the research progresses during the PhD. Research associates are usually presented with a set of tasks which the associated grant requires to be performed and the research aims are usually reasonably well defined. For both PhD student and RA, if the aim is loosely defined, then a period of initial work may be required to help further define the aim. The term 'playing' or 'undertaking pilot studies' characterises this period. The researcher tries a few different experiments (real or in-silico), looks at the methods and data, sees if there is something unexpected, works out why, and repeats this until it is clear what the results landscape is. Once this initial experimentation is finished, then the aim can be clearly set, the methods finalised to address the aim and the experiments undertaken.

For some researchers, this initial period does not come to an end and no aim is defined. Instead there is unending iteration between methods and results which can be referred to as 'activity', not 'research'. A good supervisor needs to be aware of this possibility

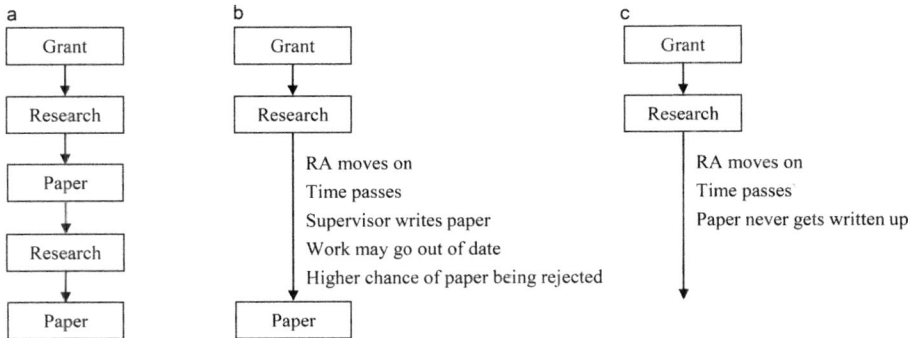

Figure 9.1 (a) Research-then-paper model of undertaking grant funded research.
(b) What can go wrong if the researcher leaves before the grant is finished.
(c) In some cases, papers are never written up

and needs to step in to break the cycle. The list above beginning 'paper reading' is activity, not research.

The above discussion has defined the research process as one in which there is an aim, some research in the form of data collection and data analysis, and a conclusion. Let us now look at where paper writing fits into this process. One model of undertaking research is illustrated in Figure 9.1a, for the case of grant funded research. The figure shows research (data collection and analysis) arising from a grant followed by publication of a paper, then more research followed by another paper and so on. This might look okay but the paper is only thought about following completion of the block of research. If the aim is defined retrospectively, then only a small percentage of the data collected is actually used which clearly means that a lot of time has been wasted collecting unnecessary data. Another common scenario is that the RA moves on after the data has been collected but before the paper has been written up (RAs will rarely wait to the end of a contract before moving on). In this case, the paper is written up slowly by supervisors (Figure 9.1b) or, in some cases, never written up (Figure 9.1c).

9.3   Paper-driven research

A relevant question to ask is when should the first paper be planned? Should this be after the data have been collected, after the grant has finished, or after the RAs have left? Clearly, all these scenarios will lead to difficulties with potentially no or low paper output as exampled in the previous paragraph. Writing the paper should be an integral part of the research, not a separate exercise. Therefore thinking about the aims of the first paper is best done as early as possible. Grant applications are often structured in work packages which can form embryo papers. Some PhD projects are

```
┌─────────────┐
│    Grant    │
└─────────────┘
       │
       ▼
┌─────────────┐
│  Research   │
│  and paper  │
└─────────────┘
       │
       ▼
┌─────────────┐
│  Research   │
│  and paper  │
└─────────────┘
       │
       ▼
┌─────────────┐
│  Research   │
│  and paper  │
└─────────────┘
       │
       ▼
┌─────────────┐
│  Research   │
│  and paper  │
└─────────────┘
```

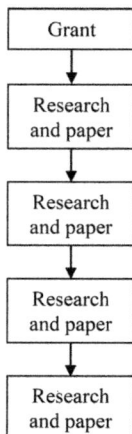

Figure 9.2 Paper-driven research; the paper writing and research proceed simultaneously

similarly structured. In the author's experience, the most effective way of undertaking research is to structure this in terms of papers to be published. Once the aim is defined, then the data collection and analysis can be undertaken while simultaneously writing the paper, and often both finish at almost the same time. The paper can be written with gaps which are filled in as the research continues. The final data collection is undertaken and the first draft of the paper appears soon after. In this way, research is structured as a continuous stream of papers with no separation between data collection/analysis and paper-writing (Figure 9.2). Those researchers who undertake this paper-driven approach usually end up with large numbers of first author papers, and their supervisors do not have to ghost-write papers on their behalf.

9.4   Structuring the paper

The remainder of this chapter will look at how a paper should be structured; further details on structuring a journal paper are provided in Chapter 13. The same rationale as above can be applied in that the structure and content of the paper should follow the research aim. Moreover, the structure and content of the paper should mirror the research itself and not be considered to be a separate exercise. Both paper and research should be organised in the following manner:

- **title** related to the main theme of the paper,

- clear **introduction** describing relevant publications, which are sufficient to justify the aim of the paper,

- **aim** clearly stated: 'the aim of this paper is …',

- **methods** relevant and clearly stated; should be sufficient to allow another lab to reproduce the results of the paper,

- just enough **results** presented to address the aim of the paper; results should be quantitative with statistical testing,

- **discussion** of limitations and relevance of the study,

- clear, short **conclusion** which addresses the aims of the paper.

In general, a minimalist approach should be taken in research, following the 'just enough' approach. For the research, just enough data should be collected to address the aim. In the paper, just enough introductory material should be presented to justify the aim of the paper, the methodology should only include material which enables the aim to be addressed, just enough results should be presented to address the aim, and the conclusion should address the aim. The researcher is advised to follow the maxim 'If in doubt cut it out' in paper writing.

Papers structured in this manner have at least a reasonable chance of being accepted by the journal. The referee of the paper will certainly look for all of the above, along with the importance of the research, the contribution to the field, whether the paper fits in with the journal scope, etc. Papers which stand a high chance of being rejected by referees have one or more of the following problems:

- do not mention key previous papers in the introduction,

- do not justify the aim,

- do not have an aim; this is the second worst fault,

- have an aim which has been addressed before or which is unimportant; this is the worst fault,

- have methods which are unable to address the aims, i.e. fatally flawed methods,

- do not discuss the limitations of the results or compare with other groups,

- do not have a conclusion (usually because there is no aim),

- put results in the methods/discussion/conclusion section,

- put methods in the results/discussion/conclusion section, etc.,

- have several aims,

- are unclear / impenetrable / overcomplicated,

- are overlong.

Many of these problems are violations of best-practice defined above. As noted in the list, the worst error is to have an aim which has already been addressed in another paper; this usually leads to immediate rejection of the paper. The second worst error is for the paper not to have any aim at all.

9.5   Summary

Research involves three elements: an aim, a conclusion and a journal paper. Research is undertaken most efficiently if the hands-on research (data collection and analysis) is performed in parallel to the paper writing. Using this paper-driven approach to research generally leads to minimal delay between finishing the research and submitting the paper, and leads to a high number of publications for the researchers without the need for supervisors to ghost-write. Research and papers are best constructed in a minimalist manner with a 'just enough' approach. Papers should be structured in a top-down manner with everything geared around addressing a clearly stated aim.

10 Project management

Peter R. Hoskins and Patrick Finlay

A project is a defined piece of work which takes place over a fixed timescale. A project team is assembled to complete the project and there is a project manager whose duty it is to deliver the work; complete, on time and on budget. The term 'project' as defined above is generic; the project can be very large such as building a nuclear power station, or it can be small such as putting in a new kitchen or building a shed. Projects come with special challenges associated with the temporary and time-limited nature of the work. The first half of this chapter discusses projects and project management in a generic sense. The second half of the chapter discusses project management of grant-funded research in universities.

10.1 Project work and routine operations

It is worth looking at a higher level as to what constitutes work and where a project fits in the workplace. The fundamental unit of the workplace is the organisation: hospital, university, company, self-employed person. Each organisation has a set of deliverables that define what the business does: treating patients, training students and publishing research, building and selling products, selling services to end-users. All these organisations have two types of activity – routine operations and projects – and both are essential to the running of the organisation.

- *Routine operations.* These are the day-to-day activities which employers undertake in order that the organisation's deliverables are delivered. Routine operations are permanent, regular, repetitive and predictable. Each organisation will require people with specialist skills in order to undertake the routine operations. The organisation will ensure that there is a match between the skills required to undertake the range of jobs and the skill-set of the workforce. Maintenance of key skills in-house is a critical issue in that the organisation could not function unless it had employees with the relevant skills. For this reason, those posts with critical key skills are usually permanent in nature, rather than time-limited. The situation is different in organisations where core skill-sets can be learnt reasonably quickly, or where there is a large supply of new employees, or where workload varies with time (e.g. seasonal work). In these cases, the employer may choose whether its main workforce is on short-term or even zero-hour (no guarantee of work) contracts, as this provides flexibility to quickly decrease or increase the size of the workforce with minimum cost to the organisation.

- *Project work.* A project is not a routine operation; it is different, concerned with the completion of a specific task. A project is time limited, with a start-date and an end-date. Usually, a group of people are brought together to undertake the work of the project. These people often do not know each other before the project starts. They may come from different parts of the same organisation, different organisations, or they may be specifically employed to undertake the work of the project with their contract ending when the project ends. The temporary nature of project work, and the fact that it is different to routine operations mean that the skill-sets required may lie outside the existing workforce. The skills required are brought together in the project team, and may be lost when the project team is disbanded.

The above division of work into 'routine operations' and 'project work' is a typical description of the different work-types in an organisation. A question which is relevant to ask is: why does an organisation need projects; or, if the core business is routine operations then what function do projects serve? Put another way, if the projects are not helping the core business (routine operations), then why are they needed at all? Three main purposes of projects can be stated.

- *To install or create new infrastructure relevant to routine operations.* This might be a new facility, new hardware or software platform, new financial accounting system, or new delivery system.

- *To expand routine operations into areas new for the organisation.* In this case, the development of these new routine operations can be treated as a project in that, by the end of the project, the new routine operations are up and running. Examples are launching a new product or a new line of business.

- *When 'routine operations' involve project work.* In many organisations, there is a project team whose job it is to undertake time-limited work which has a start-date, an end-date and a deliverable. The team may undertake several projects at the same time, with an individual member allocated to one or more projects. Builders and architects work to this model; other examples include shipbuilders and commercial RTOs (research technology organisations). In many such organisations, the project members are permanent staff, the project team mostly know each other, and the expertise is mostly maintained in-house rather than walking out of the door when the project finishes. The most obvious example of 'projects' being core business is universities where a large fraction of income is related to grant funded project work; however, in this case (as noted in other chapters), the project team consists of staff on short-term contracts, many of whom do walk out the door when the project finishes.

10.2 Phases of project management. I. Generic

This section will give a brief overview of the key steps in project management in a generic sense. Further reading on theory and practice of project management can be found in standard texts such as *Project Management for Dummies* (2013) and *A Guide to the Project Management Body of Knowledge* (2013). There is also project management software which may be bought and associated guides such as *Microsoft Project 2013 Step by Step* (2013).

10.2.1 Initiation

This is the phase where a rough idea is worked up to the point where the project aim and the anticipated outputs are clearly defined, and the potential benefit to the organisation has been defined, considered and approved. Detailed specification of the project aims, timescales and costs are absolutely key to success. There may be several layers of specification: one defining the need which has to be satisfied (why the project is being done) and another defining what technical activities are being undertaken (how the project will be tackled).

10.2.2 Planning

This phase consists of the detail of how to get from 'aims' to 'deliverables'. The end result of planning will be descriptions of all the major and minor tasks of the project; who is responsible for each task; the required equipment and consumables; the personnel required with details of the key skills for each post; a timeline of when each task begins and ends; and how each task relates to other tasks.

An important part of planning is risk management. The project plan should consider (a) a list of what things might go wrong or proceed in an unexpected manner in the project, (b) how serious each of them would be, (c) how likely they each are to happen, and (d) what mitigations can be put in place to minimise them happening and/or reduce their impact.

Following on from the discussion of risk management, an equally important part of planning is the concept of the 'critical path'. The critical path is the sequence of work packages which together have the longest overall completion time. Any delays in work packages on the critical path will delay the overall project. Part of planning involves reducing risk to a minimum for each of the critical path tasks.

The hallmark of the planning phase is the Gantt chart named after Henry Gantt who was one of the pioneers of project management in the early 20th century.

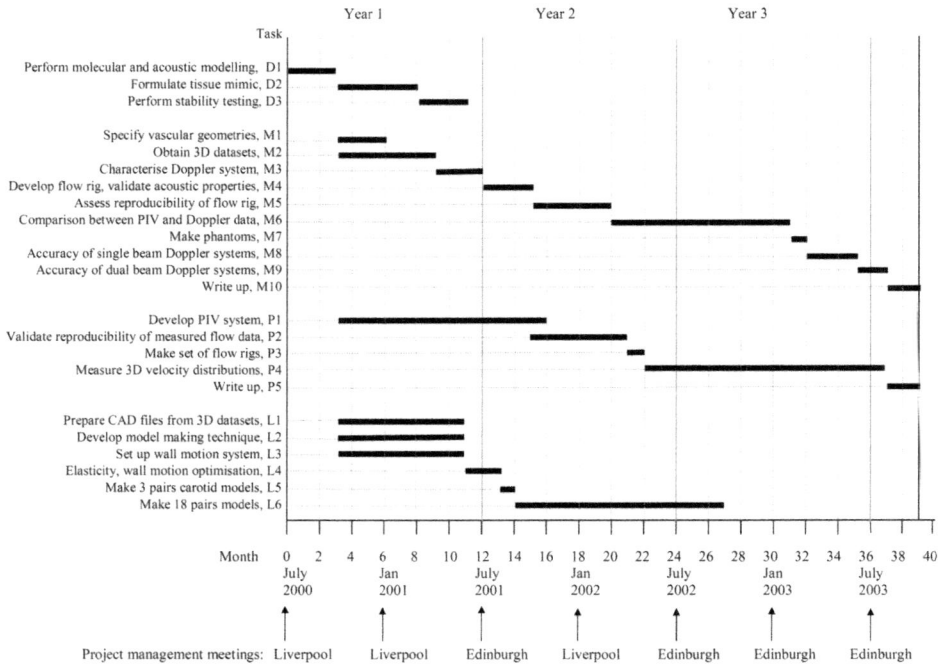

Figure 10.1 Gantt chart showing the different tasks of a grant and the associated timescales

This is a bar chart of all the key tasks with tasks on the vertical axis and time on the horizontal axis. The Gantt chart provides a visual representation of the flow of work in the project, and provides some evaluation of progress of the project in the execution phase. However, a Gantt chart by itself is often not enough: it does not show outputs or dependencies between different parts of a project or the critical path; hence, it needs to be supplemented by more detailed project planning down to the level of work packages and individual tasks within work packages. An example Gantt chart is shown in Figure 10.1 based on an EPSRC (Engineering and Physical Sciences Research Council) funded project held by one of the authors (PH) from 2000 to 2004. This was a project involving the construction of anatomical phantoms (Meagher *et al.*, 2007; Watts *et al.*, 2007). Four centres were involved in different work packages with four research associates. While the Gantt charts look beautiful, the project only bore passing resemblance to the chart for two main reasons: there was very poor progress in one arm and item D2, formulation of the tissue mimic, was vastly more complex than originally anticipated and took around 36 months, not 3 months. In the end, bits of the project ended up in other projects and studentships and related papers trickled out over the 10 years following the end of the project.

10.2.3   Execution

This phase is concerned with undertaking the main work of the project. This is usually the longest phase of a project and is where most of the project expenditure is incurred. However, this is not always the case; large infrastructure projects often take more time to plan and negotiate than to execute (e.g. HS2, Heathrow expansion).

The first part of execution involves a number of administrative tasks concerned with getting the project up to speed: hiring or deploying staff into the project, buying or deploying equipment to be used in the project, setting up lab and office space for the project team, creating a financial account which can be drawn on to pay for project costs. This all takes time and can mean that a project does not have a clean start; in turn, this will impact on the Gantt chart. Where a clean start is essential, the employer may release project funds early and take some of these setting up tasks outside of the actual timeline of the project, so that as far as possible, the main work of the project starts with everything in place on day 1.

The main part of execution involves the project team members all working on their respective tasks. Initially, the project plan is used as the basis for the activities of individual project members. As the work progresses, how far the project deviates from the plan depends on the type of project. If the project is highly predictable, in that the type of work has been done many times before, then the project plan in reality may closely resemble the Gantt chart.

In all projects, there will be variations in the training needs and work-rate of different individuals which will inevitably lead to variations in progress of the different project tasks. There will also be issues such as staff leaving or taking sick leave, equipment breaking down, and minor unexpected issues, all of which will affect the progress of the project. These issues need to be accounted for in order to keep the project on track and to ensure that the deliverables are delivered on time and on budget.

Where there are major unknowns, or where there are tasks which have never been done before, then it is likely that major changes in direction will be needed. For these projects, the project plan evolves during the project. In some cases, it may become clear that some deliverables cannot be delivered in their original form which may require negotiation with the project funder.

The later stages of execution should involve the whole project team working together to ensure that the final agreed deliverables are delivered. This is often the trickiest part of the project as all the different components of the project need to be integrated together; in medical device design, this is referred to as 'system integration'. At these closing stages, the various different teams will have all built their own subsystems and each will have been debugged individually and will function properly. However, when it comes to joining them together, all sorts of unforeseen problems usually emerge. System integration usually falls to the project manager, and it is very wise to plan for it all of the way through the project: for example, in device design, this would mean insisting on interface specifications between subsystems right at the start, and to test these in simulation as the project continues.

10.2.4   Monitoring and control

These are the systems and procedures which have been put in place to track project progress, to identify problems and to undertake corrective action to get the project back on track. The project manager is the person who takes overall responsibility for delivery of the project and is responsible for ensuring that monitoring and control take place. This aspect of project management relies on the production of information by the project team against which progress can be judged. Ideally, progress reports should be written in terms of progress of the work against the deliverables and the Gantt chart.

Project management meetings form an essential monitoring tool. For small projects involving a few people, this might involve regular weekly, monthly or 3-monthly meetings of the whole team, plus regular meetings with individual members of the team to discuss confidential issues such as performance. The team project management meeting also serves the joint purpose of a team-building exercise where everyone gets to know what everyone else is doing on the project, and where each team member learns where their own contribution fits with the wider project objectives.

For larger projects, a much more complex and detailed approach is taken with layers of people responsible for monitoring of work and reporting of this, usually via some form of formalised project management system. In addition to monitoring, decisions need to be taken about corrective action. These decisions are the ultimate responsibility of the project manager. The project manager needs to monitor and control several different types of activity including: performance of the overall team, performance of individual members of the team, project costs, timelines for work-tasks, and scope of the project.

The leadership style of the project manager is also important. If a single department or company is involved, then the project manager should have the authority to issue instructions to be followed (obviously after input from the project team). However, if it is a multi-department, multi-centre or multi-company project, then the project manager also needs to be a negotiator in that decisions on project direction will require consensus with other key people.

An essential part of monitoring and control is communication. All project team members should be aware of any changes to the work plan and timescales which affect them.

10.2.5 Closing

This final phase concerns bringing the project to a conclusion. In the last few months of the funded period of the project, the project team's efforts should be focused on the final work which needs to be undertaken to ensure delivery of the final agreed deliverables. The funding body should be provided with the deliverables along with any associated report, and should sign these off. Project staff either return to their original departments or, if they have been employed specifically for the project, their contract is ended. Lab and office space is tidied up and released back to the employer. Equipment and consumables purchased specifically for the project are reused elsewhere in the organisation, sold or disposed of.

10.3 Phases of project management. II. Research grants

This section expands on the generic phases of project management described above, giving detail relevant to research grants.

10.3.1 Initiation

Research project grant applications are written around a central aim, which is usually written in the application as 'The aim of this proposal is…'. Generation of this aim and the associated objectives constitutes 'initiation' for project grants. An aim can spring into a researcher's head fully formed: the light-bulb moment. In other instances, there is some general idea which needs developing over discussion amongst colleagues. This is more common in interdisciplinary research: one party (e.g. a bioengineer) may have an idea for a new clinical diagnostic or therapeutic procedure and approaches the other parties (clinicians, biologists) who know about the clinical area. Brainstorming is required in order to see if a viable, original and fundable project can be constructed from these initial ideas. By the end of these initial discussions, a potentially fundable project is one where:

- there is an original idea;
- there is a need for this work within the relevant community;
- this falls within the scope of a grant giving body;
- there is sufficient expertise in the team of co-investigators to undertake the research work.

If the project is aimed at an engineering funding body such as the EPSRC, the main body responsible for funding research in physics and engineering in the UK, then it should be clear that there are sufficient novel engineering challenges in the project and that there is sufficient clinical interest in the later stages in order to ensure that the work enters a clinical evaluation phase. Conversely, if the project is aimed at a medical funding body such as the MRC (Medical Research Council – the main body responsible for funding clinical trials in the UK), then it should be clear that the relevant technologies have been evaluated and are in place and that there is an unfilled clinical need which the clinical trial addresses.

During the initiation phase, it should be decided who the Principal Investigator (PI) is. Grant giving bodies expect the PI to take responsibility for delivering the grant on time and on budget. In smaller projects, the PI is usually the project manager, however, particularly on major grants, a separate project manager may be employed to manage day-to-day work.

If it is decided that a fundable project has been thought up, then the team can move on to the next stage of 'planning'.

10.3.2   Planning

The planning phase requires the project manager and team to think their way into the minute details of the work which will be performed during the project. Starting with the project aims, what are the actions that need to be undertaken first and to what purpose? Looking at things from the project deliverables end, what needs to be undertaken to steer the work towards these deliverables?

The level of detail depends on the type of project. Where there is a large body of previous experience in the tasks involved, the planning phase will specify work to a very high level of detail. This might be the case in a clinical trial where there is usually the expectation that the methodology has been worked up to an advanced state in prior research.

Research in bioengineering and medical physics often involves large unknowns. If there is no new engineering, then the research will not be of much interest to grant giving bodies such as the EPSRC who aim to fund new engineering research. In many cases, key aspects of the research of interest may never have been attempted before anywhere in the world. The logical corollary is that this research may turn out not to work at all; this is called 'high-adventure' research. Some grant giving bodies will encourage high-adventure research if there is also high reward. However, the construction of a complex project with elements of high adventure needs to be undertaken with care. A project plan should be constructed which has a balance between risk and adventure such as the following.

- Not all deliverables depend on high-risk work-tasks, i.e. there is a guarantee of some output even if the high-risk tasks don't work out.

- If an underpinning work-task (i.e. one which later work-tasks rely on) is high-risk, then there is at least one contingency which can be put in place which is lower risk.

- Pilot studies should be undertaken where possible to provide evidence that the new methodology is achievable, i.e. the risk of the work-task associated with new methodology is reduced.

The last of these bullet points will be discussed further. Pilot studies are a great idea, but they can be time consuming and where is the funding to undertake these? Sometimes these can be incorporated into a PhD studentship, or an undergraduate or MSc project, or the PI themselves may be able to perform the work. In some cases, the pilot study requires so much funding that it would need a separate grant; a grant in order to write a grant! Most grant applicants do the best they can to undertake pilot studies within their resources and accept that the incorporation of medium- to high-risk work-packages is a necessary part of the grant application.

Any project involving use of animals or applications on humans will involve ensuring that regulatory requirements are met. Some grant giving bodies will require ethics committee approval and licensing to be in place at the submission stage, others will accept that the regulatory process can be started once the funding is awarded. There will also be data confidentiality and security issues which must be addressed in the planning phase.

The end phase of the planning process is the detailed methodology and Gantt chart, and the detailed breakdown of project costs, all incorporated into the grant application, and approval from the university that the grant application can be submitted to the grant giving body.

10.3.3 Execution

If the grant is funded, then the next phase of the project is execution. The initial period of execution involves getting the project up to speed. An account is created specifically for the grant from which grant costs can be drawn. A definite start date is agreed with the grant giving body after which funds can be drawn from the grant account.

The project team needs to be put in place. Unless the project staff are named on the grant application, it is necessary to go through an appointment process involving advertising, interviewing followed by selected staff working their notice period. For non-EU, EEA or Swiss staff, a visa is necessary (noting that, during the writing of this article, the UK has voted to leave the EU making future arrangements for EU staff uncertain). Appointment of staff can lead to major delays, and in a large project there may be several researchers appointed over the first year or more of the project rather than all at the beginning. Essential equipment, especially computers, and consumables need to be bought for use in the project. Purchase of larger pieces of equipment will need to go through a tendering process.

The main part of execution involves undertaking the work of the project. The work packages were defined at the planning stage so, initially, the members of the team each start on their various work packages. There may not have been an exact match between the ideal person specification and the person employed. The individual researcher may need some training and a period of familiarisation; this period can typically be 6–12 months. The individual researchers will also have different speeds of work.

The setting up of tasks, training and familiarisation mean that most grants involving two or more researchers will not have a clean start and in practice will ramp up over the first few months, possibly even the first year or more. Even if the work goes smoothly without any unexpected technical problems, all of the above will mean that, by the end of the first year, there is a variable rate of progress across the project which will impact on the Gantt chart.

If the grant continues to run smoothly without encountering any unexpected problems, then later stages of the project will be characterised by work packages at different stages of completion, but still all heading in the direction originally anticipated. Final execution of such a project might involve the team working together to ensure that all of the work packages are finished on time. In science, this type of model might apply to a clinical trial where methodology has been through extensive development and testing even before the grant application is submitted (though this is not true of all clinical trials). By the time the clinical trial is started, all of the methodology is usually in place so that there are no unexpected issues concerning methodology. There may be other types of issues associated with recruitment, equipment breaking down, variable pace of work in different parts of the project, or staff leaving, but these should not affect the main progress of the project.

All projects will come across the unexpected requiring some replanning of work priorities. In blue-skies research, the unexpected and unknown is often the main research territory so that there should be an expectation that replanning and change of direction are necessary. In practice, often very quickly, the Gantt chart will need to be redrawn. It may become clear that a particular work package needs major modification, is much more challenging than originally anticipated or will not work at all. If possible, the project plan is shifted in order to continue to meet the project deliverables. Where it is clear that this is not possible, then the deliverables need to be changed. Changing the major or even the minor deliverables for some projects would be inconceivable, such as building a hospital. However, in science, which is a process of exploring the unknown, changing both minor and major deliverables is common. From an execution point of view, the necessity of change in direction should be expected by the project team, the project manager, and also by the funding body. Problems arise when there is a mismatch between the expectations of the project manager or funding body and the reality of undertaking the project work, which will be discussed in the next chapter.

Where major changes in direction have taken place, the later stages of execution may bear little resemblance to the original Gantt chart. The project manager, with input from the project team, will have had to undertake replanning at various stages in the project and agree revised work packages and deadlines. This replanning should ideally be aimed at meeting as many of the original deliverables as possible. In some cases, there will be new deliverables which have arisen during the course of work and the project team should have an agreed plan as to how to deliver these.

Change in project direction may require a reconfiguration of the team. The skill-set of the original team was defined according to work at the original planning stage. Change of direction, especially where certain deliverables have been abandoned or radically changed, will clearly impact on the skill-set needed. If the current team members are able and willing, then some retraining can be undertaken; however, in some cases, project posts will become redundant, meaning loss of some of the original project members and hiring of new people with the required skill-set.

The main message from the above discussion is that execution of research grants generally does not proceed according to the initial plan in the grant application. In practice, there is usually a continuous cycle between planning and execution throughout the execution phase.

10.3.4 Monitoring and control

In most grant-funded research, project management is undertaken through regular meetings of the whole team: all of the grant holders, researchers and collaborators involved in the project. Regular meetings 3–4 times a year are a common model. These meetings are typically led by the PI. Typically, each major group and researcher give a presentation of progress. There is a discussion of how the project as a whole is progressing with respect to the Gantt chart. Though not essential, it is helpful if these meetings have a focus around the publication of journal papers; these are usually the manifestation of the deliverables. Each researcher should have planned publications associated with their work and be presenting both technical progress and progress on paper writing and submission.

As noted above, these meetings also act as team-building events in which the members of the project get to know each other. These meetings are especially useful where different components of the project are in different geographic locations, a common feature of interdisciplinary research.

Outside of regular meetings of the whole group, individual work package leaders may have regular meetings with their own researchers. A common model is more regular meetings, possibly weekly, early in the grant and less regular meetings (monthly) once the grant is up and running.

While the above meetings form the backbone of project management, the real test of the project manager arises when the project does not go according to plan.

Possible problems include inadequate researcher performance, equipment breaking down, a high-risk work package turning out to be impossible to achieve, researchers threatening to leave or actually leaving at a critical point in the project, or a grant-holder using their resource to do work they are interested in rather than the work which was agreed in the grant application. Unexpected problems are, well, unexpected so can't be specified. Some of these issues and problems are discussed in more detail in the next chapter. The main point is that, once it is clear that there is a problem, then the project manager needs to act quickly and effectively to sort it out. Just sitting back thinking that the problem will go away is not an option. It is generally wise to seek the advice of others before acting. This may include communication with other grant-holders, discussing relevant issues (but not confidential issues) with the research team themselves, discussing personnel issues with Human Resources, seeking general advice from senior colleagues including the Head of Department, and discussing financial problems with your friendly contact in the finance department. It is a good idea to keep a paper trail of communication with key individuals and a record of key meetings and decisions. This is generally good project management practice. However, a good paper trail also acts as evidence that the project manager acted according to the rules of their organisation and of the grant giving body, and in the best interest of the project, in case any decisions are challenged, e.g., by the grant giving body. It is noted that completed grants are audited, sometimes years after they have finished.

Extensions of the end-date are usually permitted by the grant giving body for a variety of reasons associated with execution of the grant. The most common reason is a key researcher leaving mid-way through the grant and the replacement researcher having to get up to speed. Other reasons include: disasters such as damage of labs and equipment (e.g. by fire, water or vandalism); medium- to long-term leave of a researcher due to illness; failure of delivery of key equipment, software or services from an external provider. In general, these extensions come with no extra funding, but the PI may be allowed to make the case for diverting funds from other parts of the grant where there has been an underspend, which allows key researchers to be employed to the end of the grant.

10.3.5   Closing

In the last 6 months of a project, it should be clear what is achievable and what is not within the remaining period of grant funding. By this stage, there should be no more changes of direction. The main issue is finishing the final agreed work packages and attempting to deliver the final set of deliverables. Each member of the team should have a very good idea what they have to do in parallel to what journal papers they are writing up. During this period, some of the research team leave for the next post, not waiting until the end of the grant before they leave. The period of closing a grant in terms of final delivery of deliverables can be challenging as the resource to undertake the work gradually diminishes.

After the grant end-date arrives, then no more money can be drawn from the grant account. The PI needs to keep an eye on the diminishing resources in the grant account during the final months and ensure that all bills have been paid. It is not fun to receive a £20k unpaid bill after the account is closed. Working closely with the local finance administrator is essential during this period.

What happens to remaining consumables and equipment after the grant has ended depends on local practice. The traditional approach is that a PI has their own lab, so that after the grant has ended the equipment just remains in the lab ready for the next grant if and when this arrives, with the equipment and consumables all belonging to the PI. In practice, however, it is actually the university that owns the equipment, not the PI. With the pressure on space in universities, the more modern approach is that lab space is allocated depending on grant funding. When the grant finishes then the benches and offices are cleared. Any equipment not needed is reallocated and if necessary sold or disposed of. In the UK, the EPSRC insists that the university contributes 50% funding to major pieces of equipment so that these become shared resources which continue to get used after the grant has finished rather than the equipment ending up unused or skipped.

Once the grant is finished and the account closed, then the PI has a fixed time of 3–6 months to complete the final report for the grant giving body. Failure to complete the report can lead to a fraction of the money (e.g. the final 3 months of funding) being withheld by the grant funder, so completion of the final report is essential.

In practice, closing of a grant occurs in phases; closing the grant account is the first major phase, submitting the final report the next. However, researchers who were employed on the project and the PI will continue to write up papers from the grant long after it has finished. In this sense, most projects wind down gradually in the same way that they started up gradually.

10.4   Summary

Project management is an essential part of grant-funded research. The grant PI becomes by default the project manager, though for larger projects, it is common to delegate some project manager functions. The initiation and planning phases become crystallised in the grant application. The execution phase, especially in larger multi-researcher grants, winds up gradually. Research in Medical Physics and Bioengineering is usually associated with large unknowns so that, during the execution phase, there is a continuous iteration between planning and execution, in some cases with signficant changes in minor and major deliverables. Project management meetings form the backbone of project management; however, the project manager must also create communication channels with the project team members to pick up problems as these arise. The project manager must also act quickly and effectively to sort out these problems seeking advice as necessary. Closing the project is a gradual process, often with the PI and researchers writing up papers from the grant long after the grant funding has ended.

References

Chatfield, C. and Johnson, T. (2013) *Microsoft Project 2013 Step by Step*. Microsoft Press, Redmond, WA.

Meagher, S., Poepping, T.L., Ramnarine, K.V., Black, R.A. and Hoskins, P.R. (2007) Anatomical flow phantoms of the nonplanar carotid bifurcation, part II: experimental validation with Doppler ultrasound. *Ultrasound in Medicine and Biology*, Vol 33, Issue 2, 303–310.

Portny, S.E. (2013) *Project Management for Dummies*. 4th edition. Wiley.

Project Management Institute, Inc (2013) *A Guide to the Project Management Body of Knowledge (Pmbok Guide)*. 5th edition. Project Management Institute, PA.

Watts, D.M., Sutcliffe, C.J., Morgan, R.H., Meagher, S., Wardlaw, J., Connell, M., Bastin, M.E., Marshall, I., Ramnarine, K.V., Hoskins, P.R. and Black, R.A. (2007) Anatomical flow phantoms of the nonplanar carotid bifurcation, part I: computer-aided design and fabrication. *Ultrasound in Medicine and Biology*, Vol 33, Issue 2, 296–302.

11 Practical issues in project management of research grants

Peter R. Hoskins

11.1 Introduction

The aim of this chapter is to discuss practical issues in grant management with respect to the different parties involved. As with previous chapters, the emphasis is on grants relevant to physicists and engineers working on biomedical research. As noted in the previous chapter, the principal investigator (PI) is responsible for delivery of the grant on time and on budget. For small and medium sized grants, the PI is usually the project manager. However, some project management functions, especially for larger grants, may be delegated to a project coordinator.

Many of the practical issues in grant management are associated with the theme of 'meeting expectations'. A model of grant management might imagine a perfect world in which grant holders are dedicated to delivering the grant on time and on budget and provide suitable supervisory input as needed, that all research associates (RAs) are similarly dedicated and all stay to the last funded day of the grant to ensure that the work is delivered in full, that the quality of work produced by all RAs is at a suitably high level, that collaborators such as industrial partners will deliver the support to which they agreed in the grant application, that the grant giving body understands that where research has never been done before then it may not actually work at all, and that the PI's department understands the nature of physics and engineering research and is generally supportive to both grant holders and the RAs. If these are the expectations of the project manager, then they will quickly find out that one or more are untrue. In this context, the PI needs to match their expectations with the reality of how a grant runs in practice. Only when there is a match between reality and the PI's expectations will they be an effective project manager.

These examples are drawn from a wide range of grants and individuals in different institutions. The examples deliberately discuss an imperfect world and the reader should not think that all grants have all the issues listed below. However, some of the more extreme examples of negative practice given do occur and it is better that the prospective project manager is aware of these from the beginning rather than discovering them as they go along.

11.2   Grant giving body

Before the era of interdisciplinary research, the division between physics/engineering research and biomedical research was clear: physics and engineering grant applications went to a physics and engineering body (in the UK, this is mainly EPSRC); biomedical applications to a biomedical body (in the UK, examples are MRC, BHF and Wellcome). Physics and engineering grants were reviewed by physicists and engineers who understood practical issues of research in these fields, and the grant giving body understood how research in these fields was conducted, including the necessity to change research direction during the grant if some methodology did not work. In other words, there was a match of expectations between all of the main parties: the grant applicants/holders, the referees and the grant giving body.

Interdisciplinary research involving engineering/physics and biology/medicine has been growing rapidly since the 1990s. In 2016, the area of bioengineering is the most rapidly growing field of engineering in the world. In terms of interdisciplinary grant applications, it used to be common for a grant to fall between grant giving bodies; not enough new engineering for a body such as EPSRC but too much for a body such as MRC; too clinically oriented for EPSRC but too much reliance on unproven technical methodology for MRC. Even now, when interdisciplinary research is common and where there is working between grant giving bodies, grants may end up falling between the two bodies. The advice given here to a PI of an interdisciplinary grant application is to configure the aims and methods to suit the grant giving body. Discussion with one of the body's programme managers certainly helps. In some cases, it may be necessary to go through an iterative process of submitting the grant to one body, modifying after rejection, then submitting to another until there is a match between the details of the grant and the remit of the grant giving body. In 2015, a colleague in Edinburgh went through this process with three submissions over 2 years before he was eventually awarded £5 million.

There are also issues during other phases of a grant. The main issue concerns the need to change research direction for physics and engineering grants. In physics and engineering grants, the methods section is often written as a first best-guess, with the acceptance that this will evolve during the grant. Problems can arise if there is funding of this type of project via a grant giving body that normally funds medical research such as clinical trials. The programme manager for the grant giving body may take the opinion that any deviation from the agreed plan is a violation of the agreed contract and may challenge both the change of direction and also the purchase of any equipment that has been bought associated with the new methodology. This may arise during the execution phase or at the final report stage, even resulting in demands for return of some of the grant funding. For some grant giving bodies, there is considerable flexibility built into the use of funds. Others may allow flexibility if a good case is made and permission is sought. In either case, the PI is advised to read very carefully the terms and conditions of the grant which will include advice on flexibility in the use of the agreed funding. If it is clear that there is insufficient flexibility for a physics or engineering project, then the PI is advised not to apply to that particular body.

11.3   Host department

Physics and engineering grants are characterised by large amounts of technical methodology development and publication of this methodology in field-specific journals. The host department needs to be supportive of these activities. This might seem a blindingly obvious statement to make, and something which an academic medical physics and medical engineering department or a bioengineering department should do. However, increasingly, engineers and physicists are based in biomedical centres. These centres are characterised by hypothesis driven biological or clinical research and publication in high impact biomedical journals. This has an influence on the management of a physics or engineering research grant, with relevant issues discussed in the remainder of this section.

11.3.1   Journal publication

Research involving physics and engineering applied to medicine involves two main phases: a methodology development phase and then a phase involving application in humans (both volunteers and patients). The first phase usually involves publication in field-specific technical journals, while the second phase involves publication in clinical journals. The impact factor of the best technical journals is usually in the range 2–4 (e.g. *Magnetic Resonance in Medicine* 3.8, *Journal of Biomechanics* 2.5, *Ultrasound in Medicine and Biology* 2.1), whereas the impact factor of the best medical journals is in the range 10–60 (*Circulation* 14, *Lancet* 45, *New England Journal of Medicine* 59). Journal impact factor is increasingly used as a marker of publication quality in its own right, even when this approach is constantly criticised as not being the intent of the impact factor. This is problematic when a Centre uses impact factor thresholds, in line with submission in one of the regular research evaluation exercises, so that only publication in a journal with an impact factor greater than the threshold is considered 'high quality'. For example, a threshold impact factor of 4 rules out most of the top technical journals. There may be advice to the PI to stop publishing in 'low impact' journals, to stop doing methodology development and concentrate on off-the-shelf technology, and even to change field to an area whose journals do have high impact factors (*Journal of Biophotonics* 4.2, *Journal of Nuclear Medicine* 6.2). This latter advice would involve changing technical fields, which, even in the days where interdisciplinary research is common, very few researchers manage due to the need to learn a wide range of completely new technical skill sets and knowledge. All of this advice is incompatible with managing and delivering a physics or engineering research project or with developing the careers of the PI, the research staff and any associated PhD students.

A good clinical department should encourage the career developments of its engineers and physicists, including supporting grants which contain large amounts of methodology development and publication of first author papers in field-specific technical journals.

11.3.2 Critical mass

While a mixed environment of physics/engineering and biomedical researchers looks ideal for interdisciplinary research, problems arise if the number of engineers and physicists falls below a critical amount. For example, a new researcher learning how to use a complex technology such as computational modelling software best does this if there are a few peers located nearby who can help. In the example given, this would be 2–3 RAs or PhD students who have experience in computational modelling. This peer-peer support is essential for RAs and students to make good progress. If the physics and engineering research team is small, there can be considerable issues of professional isolation. Mixed centres work best when there are critical masses of biomedical researchers and physics/engineering researchers.

11.3.3 Local facilities and local support

The performance of engineering and physics technical work should be undertaken in labs which are suitable, and with appropriate technical support. While general labs consisting of bench space and storage cupboards may be adapted for engineering or physics work, some work may require access to specialist facilities such as mechanical and electrical workshops, involve the use of hazardous chemicals and materials such as heavy metals, and involve a level of computing which is more commonly available in an engineering department than in a clinical centre.

There should also be appropriate technical support for health and safety and IT. While this might be available in an engineering department, most biomedical centres concentrate on support for biology and clinical trials and do not have the detailed knowledge required for physics and engineering projects. In particular, the responsibility for health and safety lies with the PI, not with the Health and Safety advisor, and the PI has to ensure that the local working environment is acceptable and that all relevant safety procedures are in place and that safe practice is followed. In some cases, work (e.g. involving hazardous chemicals) may need to be undertaken elsewhere in the university where there are suitable facilities – these issues are best identified at the grant application stage. Computing support can be a challenge when working in a clinical centre. While the IT support team in a physics or engineering department will be used to supporting a wide range of software and hardware platforms including adapting the computer ('taking the covers off'), most IT support in a clinical centre is for basic computer use involving document preparation, email and web browsing. There probably will not be the level of IT support which is needed, forcing the PI to either do this within their own team or to try to access support elsewhere; again, these issues are best identified and included at the grant application phase.

11.3.4 Fitting in

If you are a physicist or engineer in a biomedical centre, then it is likely that the vast majority (95% plus) of academics in the centre will be biologists or clinicians. Consequently, virtually all of the talk will be of biology and medicine, as will the posters on the wall, the contents of the seminar series, and as noted above, the expectations of the centre. In terms of delivering a physics or engineering research grant, this leads to issues of fitting in. It is very likely that biomedical colleagues will not understand the practice of engineering and physics. In some cases, there will be a willingness to learn; this is obviously the best scenario and from this can arise genuine collaboration and the breaking of new ground, something all researchers want. There must be a willingness of both parties to engage with each other's areas and this takes time, often many months or years. Trust can be built through small collaborative projects such as final year or MSc projects and PhD studentships. Certainly if you have a grant as PI your biomedical colleagues should understand the practices of physics and engineering research, and especially the differences from their own area. If genuine collaboration between engineering/physics and biology/medicine is supported by the head of centre, then this is the best situation, and this should include some active measures to allow the physicists and engineers to fit in. In addition to encouraging methodology development and publication in field specific engineering/physics journals, this might also include encouraging physicists/engineers to be on key committees, a willingness of senior staff and support staff to help solve issues relevant to physicists and engineers, and above all, celebration of the success of engineers and physicists in the centre through the usual mechanisms (all-staff emails pointing out success, items in the centre newsletter, invitations to the seminar series, etc.).

If, however, there is no effort made to help physicists and engineers fit in, then this can be challenging. This usually runs in parallel to clinical centres who carry on with 'business as usual' rather than embracing the opportunities which physicists and engineers may bring. Physicists and engineers who are PIs may be seen mainly as providing research support rather than as equal partners. This can lead to the 'second class citizen' phenomenon. There can be requests from biomedical staff for RAs and PhD students to hand over the data so that the biomedical researcher can write 'their' paper, even though the idea was originated by the engineer or physicist and the RA/PhD student did all of the work. In presentations, biomedical staff may mention the technology that 'they' have developed with no mention of the physics and engineering team who actually did the work. The centre seminar series will be configured around clinically led or biology led themes. Requests to speak may be accompanied by 'and just skip over the methodology and go straight to the applications'. Finally, what constitutes success is seen in terms of the main themes of the centre which are biomedical and involve high impact journals. In this environment, the core methodology work and journal publications of the physics and engineering team are seen as being 'not us' and second class. These issues will impact on a daily basis on the progress of an engineering or physics grant with a biomedical focus, affecting PI and researcher morale, and ultimately on delivery of the grant. If there is no support for resolving these issues, then the PI should ask themselves whether they are best placed in a more supportive environment such as a bioengineering department. Physically relocating the whole group is possible and best done after the grant has been awarded and before the work has started – as usual in academia, funded grants are good currency when it comes to moving centres or institutions. If for whatever reason this is not possible, then practical issues which can help involve the PI and their researchers having associate membership of the local bioengineering department. This may involve attending lectures and access to a peer group. Having dual bases (i.e. the RAs have a base in the biomedical centre and a base in bioengineering or medical physics) is possible but not widely used due to the doubling of office space which this requires, and this does not help the core issue which is that the physics and engineering research is based in the wrong place.

If the research centre has a recognition of the different needs of biomedical and physics/engineering communities, then this represents the ideal environment for delivery of a physics and engineering grant with a biomedical focus. There has been some progress in this area in some universities but this is an evolving area and there remains a huge amount to do to facilitate proper joint working of biomedical and physics/engineering researchers.

11.4   Co-investigators

Multi-department and multi-institution grants are common in bioengineering and medical physics research. Multi-centre grants funded by the EU typically have 5–20 partners. Most co-investigators (co-Is) will do their best to deliver the work that has been agreed in the grant application. However, there is a small minority who do not operate in this manner. Usually, such a co-I has a problem in that they need money to fund their research, and the grant then represents a solution to that problem. The co-I then happily continues with their own research agenda which might bear only passing resemblance to the work agreed in the grant application. In a tightly structured grant application, the above behaviour is a disaster as a major component of the grant is making no progress. Sorting out this type of behaviour is challenging, requires evidence of lack of progress over many months or even years and will almost certainly require intervention at a higher level such as Human Resources. In practice, such issues can be difficult to resolve satisfactorily. Fortunately such attitudes, which lead to no useable output at all from a co-I, are becoming less common. However, the PI of a large multidisciplinary grant must expect different rates of progress in the different centres, in some cases with very slow progress. In these cases, it may be possible to divert funds from elsewhere in the grant to help undertake the tasks which have fallen behind. It may be possible to find other resources from outside the grant. Commonly one or more PhD students might be attached to a grant. Involvement in a major grant represents a superb opportunity for the student and the work of the student can be folded into the final report. Care needs to be taken as PhD students are supposed to be in addition to the main work of the grant. However, in most cases, the student can undertake their PhD in a satisfactory manner as well as contributing to the progress of the grant.

11.5   Industrial partners

The involvement of an industrial partner is often seen as being highly desirable in a bioengineering or medical physics grant. The industrial partner agrees their involvement at the grant application stage, such as provision of prototype equipment, access to specialist software and provision of engineering staff time. The problem arises if the company has a major change in its priorities during the execution phase of the grant. This may happen after amalgamation with a larger company, or during a financially lean period such as might occur following a global financial crash. In this reprioritisation, the industrial partner may no longer be able to provide the support to the grant which was originally agreed. If a key part of the grant depends on this support, then this is problematic for progress. When it is clear that there is a problem with the industrial support, then the project manager needs to look at what options there are for replanning the work. Options include bringing some of the work in-house or sub-contracting elsewhere. Decisions need to be taken as to what deliverables need to be abandoned or changed. This type of issue should be raised with the grant giving body to forewarn them of any change of project direction, and also as ammunition for asking for an extension nearer the deadline.

At the planning stage, the PI should be thinking ahead as to what would happen if the industrial partner ended up having to reduce their effort on the project. One way of reducing risk is to build in industrial support as early as possible in the project, including the provision of any prototype equipment or software. The company may be able to work on the prototypes between the grant being awarded and the grant starting, so that the equipment and software are available on day 1 of the project.

11.6   Research associates

The RAs are the most important part of the grant as they do all of the work. It is therefore essential to employ good RAs and look after them in terms of providing them with appropriate support.

Interviews may normally concentrate on the technical abilities of the RA; will the RA be able to undertake the technical programme of work of the grant? However, the RA is also expected to write the journal papers. If an RA has very high technical skills but a very poor record of first-author publication, then this is not a good recipe for getting papers published. When the RA leaves without publishing, this task then falls to the PI who must ghost-write papers in order that there is something to include in the final report. This is time consuming for the PI and sometimes is not possible as the documentation is insufficient. In the person-specification criteria for interview, some PIs decide that a recent (last 2–3 years) good track record of first author publication is essential, even more important than previous experience in the field.

Only around 9% of RAs will gain a permanent academic post. So an RA is not a career post in the same way that, say, an NHS medical physics post is. Most RAs come with the expectation that this is a temporary post which serves their current purpose, before at some point they move on to another post, possibly one associated with a longer-term career. The purpose of the RA post is obviously summarised in their employment contract in the same way as for everyone else; they get paid and in return, do the work required by the post. In addition, the RAs will have their own reasons, apart from getting paid, for taking the post. Examples might be to build an academic CV in order to be competitive for a research fellowship including evidence of independent research, to gain skills and experience useful outside of academia in a future career, to live in a nice city, and so on. The PI needs to be mindful of these needs and should expect that RAs may want to use some of their paid time to develop activities unrelated directly to the grant. Universities encourage this and so the PI also needs to encourage this. Those PIs who expect absolute 100% commitment from their RAs on the grant application are likely to find a difficult time in managing their RAs.

The other aspect of the temporary nature of the RA post comes at the other end of the grant in that RAs may not remain to the very end. RAs will often leave sometime between half way and the end of the grant for their next post. The final stages of a grant can be challenging; just at the point where all effort needs to be made to finish the grant, then this is the stage where the RAs are likely to leave. Replacement RAs can be appointed. If there is no-one locally available to take over, this means hiring in the usual way which is challenging when the post length is 6–18 months. With the time required for training and familiarisation, there may end up being very little productive effort from the replacement RA. Seeking an extension when key RAs leave is vital, as is making savings elsewhere in the grant in order to pay for the RAs' salary during the extension. The PI can do very little to plan for RAs leaving and just has to deal with this when it happens.

11.7   University support departments

Departments such as finance and human resources can provide invaluable advice on issues which arise during the grant and it is better to seek advice more often and sooner rather than less often and later. There is just one issue which is worth raising which concerns the financial statements on the current state of the grant. These are retrospective in that they show what has been spent to date, whereas the PI needs prospective information on how much money will be left in the grant when it ends based on current trends. This prospective projection allows the PI to do some planning about how any remaining money can be spent. Some finance departments offer this prospective financial planning as a routine. If this is not provided, it is worth working with the local finance administrator to develop this tool.

11.8   The PI

Finally, there is the PI. Does the PI have the general characteristics that make a good project manager or are their grants going to be associated with chaos and confusion? There are some general characteristics which it could be argued are signs of future project management expertise. If you are considering becoming a PI or are a PI, then ask yourself the following questions.

- *Responsiveness.* Do you have hundreds or even thousands of unanswered emails in your in-tray and regularly do not reply to key emails or, do you clear your in-tray and find time to reply to emails at least several times per week? If the former, then this is not a good sign that you are going to respond in a timely fashion to emails from your RAs or co-investigators who will all expect you to be answering emails promptly.

- *Timekeeping.* Are you regularly late for meetings, or do you not turn up at all? Even if you think that you are on time for meetings, do you get angry people asking you where you were yesterday as they waited at the agreed time for an hour and you didn't turn up as you had forgotten and gone to Paris? Do you regularly fail to meet deadlines for reports/chapters/papers and have to be pushed endlessly by other people for your contribution? If yes to any of the last three questions, then this is an extremely poor sign that you will be a good project manager. Good time management is essential in a project manager.

- *Sorting out problems.* When someone tries to explain that there is a problem that you need to fix, do you: i) deny categorically that there is a problem, ii) if pushed, accept that there is a problem but insist that it's not your fault or your responsibility to fix, iii) if further pushed, accept that there is a problem and that it is your fault and your responsibility but say that you can't/won't do anything about it? Disaster. In grant management, the ability to identify and deal with problems is entirely dependent on admitting that these problems exist, being prepared to accept your role in them, and actually confronting the problems head-on even though this might be uncomfortable for you.

Some PIs in practice will lack the essential skills of a project manager such as good time management, communication and an ability to sort out problems. Their behaviour may only be slightly modified by the experience of managing grants, and their grants may be accompanied by varying degrees of chaos such as the following.

- *Lack of progress.* Different parts of the project proceeding in opposite directions; critical issues affecting progress not being picked up and hence not being resolved; too much effort spent on non-essential tasks, too little effort spent on critical tasks; after much effort, co-investigators give up trying to seek direction and end up doing what they want in a manner unrelated to the rest of the project; key RAs who have been trying to raise issues without success also end up doing what they want in a manner unrelated to the rest of the project, or they give up and leave.

- *Lack of responsibility.* We have probably all met the manager who seems never to take responsibility for anything that goes wrong and to whom nothing sticks. The phrase 'teflon coated' is used for good reason. The boss shirks responsibility and both responsibility and blame bounce off the boss onto someone else. Such a working culture tends to be an unhappy one where people may copy the boss's behaviour in an attempt not to have the blame fall on them. This type of culture may be associated with a boss such as a PI who refuses to take responsibility for anything that happens in their grant. If the PI lacks the ability to recognise or admit that there are problems, then the PI is also not going to be effective in sorting out problems when they arise.

Grants can and often do struggle through without proper project management. Quite a lot of the project management functions can be delegated to a project coordinator. This is essential for very large projects and for PIs who are far too busy or important to do the day-to-day tasks of project management. Where there is no project coordinator, then some of the project management can end up being distributed amongst the other grant holders and even the RAs. Project management meetings are replaced by ongoing updates between group members who are constantly in touch with each other, especially if they work in the same geographic location. The PI might be good at writing in which case the grant application and the final report are well written. The delivered grant looks good enough from the outside and the final report receives a good enough mark justifying to everyone that the grant has been properly managed, but with no indication of the chaos on the inside, and with no learning on the part of the PI. On the other hand, the grant might not proceed smoothly leading to a poor final report which is not good for the PI or their university.

Increasingly, universities are developing a more professional approach to grant management and very poor grant management is less common and increasingly not tolerated. If the PI is responsive, meets deadlines, can take responsibility for their own mistakes and help sort out these and any other problems then, with suitable training and some on-the-job learning, they probably will become a good grant manager, and all of their colleagues will thank them for this.

At the beginning of this section, a number of questions were asked which provided a litmus test of potential project management ability. In a similar manner, there are signs of practice which may provide an indicator of whether project management is being done well or done poorly. Researchers may wish to examine their own projects (and equally those of others) to see which category they fall in. The list immediately below provides signs which may indicate poor project management:

- Emails to the PI go unanswered.

- No regular meetings involving all grant-holders and researchers.

- Obvious attempts by one or more researchers to play the role of project manager.

- Project researchers spend large amounts of time doing work unrelated to the grant.

- High turnover of project staff.

- The grant is clearly being used to finish off a previous grant.

- Overspend: there are instances of spending all of the money in the first year of a 3-year grant.

- PI can never be found.

- Statements denying the existence of problems: 'we do not have any problems here'.

- Publications from researchers bear no resemblance to the work of the grant and/or are not reported in the grant final report.

- There are no publications.

Signs which probably indicate good project management:

- PI is responsive.

- Regular meetings involving all grant-holders and researchers.

- Project researchers know what they are doing and why, and how it relates to the overall direction of the grant.

- The spend is appropriate and there is a forward-projection of spending to the end of the grant.

- Steps are taken to redirect the project as needed.

- The PI keeps the funder informed of relevant issues.

- Generally, the PI keeps a good paper-trail of key events which affect delivery.

- The PI sorts out issues as they arise.

- The PI makes an attempt to encourage publication of final papers including ghost-writing.

- The final report is well written and delivered on time.

- Most of the work of the grant ends up being written up as journal papers.

11.9 Summary

Research grants almost never run to plan and the job of the grant manager is to both pre-empt and sort out problems as they arise and ensure that as many as possible of the final grant deliverables are delivered on time and on budget. A good grant manager will have key skills such as good time-keeping, good communication, responsiveness and an ability to identify and sort out problems (including ones they have helped create). Key to success is the host department which must be supportive of methodology development and of publication of this work in the best field-specific technical journals.

12 Publication and peer review

Richard A. Black

12.1 Introduction

Research has been defined as 'a process of investigation leading to new insights *effectively shared*' (HEFCE, 2009). As scientists, physicists and engineers who work in an academic environment, we are mandated to disseminate the results of our activities to the wider scientific and clinical research communities, but also to bring to the attention of the wider public the work we do on behalf of publicly funded research bodies and charitable organisations. It follows that the generation of new knowledge goes hand-in-hand with communicating the results of research. In what has become an increasingly competitive environment for researchers, universities are expected to demonstrate the impact of their work for the betterment of society through the generation of new knowledge, and its effective transfer and exchange with industry, and how this has led to improvements in healthcare (i.e., research with impact). The dissemination of this knowledge by traditional means such as publication, via the scientific and clinical literature, and the wider media in general, remains an important contribution to this end, since the significance of many new findings may not be apparent at the time of publication, either to the authors' immediate community of practice or society at large. Writing is an essential part of that process, and is as important as any list of skills and techniques that one might expect to find in an academic CV. Arguably, the work of the academic is not simply to get that new knowledge into the public domain; it is also about ensuring that those findings reach and are read by the widest possible audience; hence, the increasing emphasis placed on citation metrics as a measure of research impact.

The quality of the research base within the UK is determined through periodic assessment exercises, conducted every 5 years or so, most recently through a process known as the Research Excellence Framework, or REF, which undertakes to assess submissions made by higher education institutions. The criteria vary according to the individual research disciplines, but the assessment is based largely on the quality of research outputs, the research environment and indicators of esteem, and evidence of impact. Dating from the mid-1980s, these exercises have evolved from being a periodic research audit to one that has major ramifications for the funding of higher education establishments in the UK, with universities already gearing up for the next exercise in 2020–21 even though the criteria have yet to be made known in detail.

The purpose of these exercises is to 'benchmark quality against international standards and to drive the Council's [Higher Education Funding Council for England (HEFCE)] funding for research', and to 'provide a basis for distributing funding primarily by reference to research excellence'. Increasingly, those measures of research excellence include bibliometric indices that are intended to capture how well a particular publication has been received within the research community. Until recently, much emphasis has been placed on where the work is published rather than how often that particular work has been cited, a journal's impact factor (IF) having been viewed by many as the primary indicator of quality. Journal impact factors are derived from an analysis of the data held on the ISI Web of Science database over a 2-year window, a journal's IF being the average number of times articles from that journal published in the past 2 years have been cited in the current year (Journal Citation Reports™).

In reality, of course, publication in a high-impact journal does not guarantee that an article will attract the number of citations that a particular journal's impact factor would suggest. Moreover, the rate at which a given publication attracts citations varies according to article type, and may not coincide with the impact factor citation window. Arguably, the 5-year impact factor is a better indicator of a journal's standing than the traditional IF metric. Other journal metrics include: Impact Per Publication (IPP), which represents the average number of citations received in a particular year by papers published in that journal during the three preceding years; and the Source-Normalised Impact per Paper (SNIP), where citations are weighted according to discipline. Just as with journal rankings, indices are used to compare individual researchers. One such index, the h-index, was developed in 2005 by Jorge Hirsch, a physicist at the University of California in San Diego (Hirsch, 2005). Hirsch's aim was to capture the impact and volume of an individual's research output in a single index. The measure is simple: a scientist with an h-index of say 20 has published 20 articles, each with at least 20 citations, the remaining papers having fewer citations. That the worth of one's entire research output to date can be expressed by a single index is the ultimate in reductionism, but it does give an indication of the impact of an individual's research portfolio.

Given the strategic importance that publications have for individual researchers as much as their employers, it is fitting that a chapter on journal publication should be included in a book directed at students, academics and clinical scientists who undertake research in the university and/or hospital setting, whether fundamental, applied or clinical in nature. Nevertheless, it is surprising how this skill is often overlooked. Writing for publication should not be an afterthought, or left until the work is completed and/or a thesis has been submitted; on the contrary, thesis writing is undoubtedly easier if the work has been written up for publication. Arguably, the process of drafting and submitting a paper for peer review may help to identify problems while the work is in progress, when there is still the opportunity to address them.

The sections that follow are based on a series of workshop presentations I have delivered to students of my university and at conferences. I aim to offer practical advice and guidance on best practices when preparing manuscripts for submission to peer-reviewed journals based on my experiences as an editor of the Institute of Physics and Engineering in Medicine (IPEM) journal *Medical Engineering & Physics* (MEP). While the quality of the science being reported in any manuscript for the most part will determine the eventual outcome, the advice below may help to smooth its passage through the peer-review process, and increase the chances that it will be cited.

12.2   Planning for publication

Many grant-awarding bodies require applicants to make clear how they will disseminate the results of their research. Indeed, researchers who undertake publically-funded research in the UK are mandated to make available the results of their research to their immediate research community on a publicly accessible repository, or via publication, within a certain timeframe. Before work on a given research project has started, it is sensible to revisit those plans, and agree how the individual work packages might translate into one or more publications. While this may seem odd before any results have been generated, it should be possible to envisage what experiments are involved, the form the results will take, and how those datasets will be analysed and presented. Arguably, the best research papers are based on one or more hypotheses that the investigators set out to prove or disprove. Hypothesis-driven research is likely to be received more favourably by the scientific community than reports of serendipitous, incremental research.

Authors who have published widely will know that they may be required to make a series of disclosures about the work reported in the manuscript: declaring any conflict of interest, perceived or actual, acknowledging the source(s) of funding, etc. For work involving human subjects, including volunteers, authors must seek prior approval from the appropriate review body in order to ensure compliance on matters of ethics. In the case of studies requiring access to patients or patient data, this will involve submitting an application to the Health Research Authority via the Integrated Research Application System (IRAS), a time-consuming process at best, for which planning is essential, since it can take several months to apply for and get approval – a major concern when working on a time-limited grant or studentship.

Other project management issues to consider include who should do what and when, and how each individual's contribution is to be recognised. Many journals require authors to include a statement on the contributions made by each author as part of their submission process, yet it is surprising how often authorship disputes arise. It is important to manage expectations from the outset by agreeing the authorship of each publication, i.e., whose name is to appear on each publication and in which order. In cases where two individuals, usually the first and second authors, contribute equally to the execution of study, this should be recognised in an explicit statement in the paper; the remaining authors follow, with the group leader or principal investigator's name usually appearing last. Any one of the authors may take responsibility for managing the process of submission (referred to as the corresponding author) and communicating to the other authors any feedback from the reviewers and the editor throughout the process of peer review.

The authors should consider where best to publish their work, and what article type is most appropriate for their study. A preliminary investigation may be better suited for consideration as a short communication rather than a full-length article, whereas a short, methods-based paper may be more suited for consideration as a technical note. These categories should not be viewed as inferior, since all have the potential to attract citations and are not distinguished in the CV or by search engines such as Web of Science® (formerly Web of Knowledge). As stated previously, the journal with the highest impact factor may not be the most appropriate; rather, the journal that best represents your own community of practice is where the work is most likely to be read and appreciated, and eventually cited.

12.3   Preparation of your manuscript

My academic career began before the advent of desktop publishing, when word processors, spreadsheets, and graphical software became the norm. Since that time, the art of publication has changed almost beyond recognition (yes, I am old enough to remember the technique of applying Greek and mathematical symbols to line drawings by hand; I find it reassuring that the company Letraset® (www.letraset. com) still exists!). Fortunately, drafting, typesetting, generating drawings and figures, etc., have evolved from being a relatively laborious activity involving a degree of manual dexterity, into a digital, computer-based one, the speed of which is determined to an extent by the author's keyboard skills. Few academics these days have personal assistants whose job it is to type up handwritten notes. Almost without exception, the preparation of a manuscript and its submission to the publisher for peer review are done on line, via email and websites administered by the publisher. As a result, the process is managed largely by the authors themselves, with little input from copy editors, typesetters and the like; consequently, authors are responsible to a far greater extent than in the past for the appearance and presentation of their output. Fortunately, most publishers provide detailed guidance on formatting requirements for manuscripts submitted to their journals; some even provide a template document in MS Word or LaTeX for the author to complete. Authors are advised to consult and follow those instructions carefully, as they will contain information pertinent to the preparation of a manuscript, such as the accepted formats for citation, artwork, file formats, etc., which may well save time in the long run.

Increasingly, authors are encouraged to make use of online supplementary information to complement the traditional Portable Document Format (PDF) or print versions of an article, as on ScienceDirect®, for example, which provides direct links to the supplementary files from the article's web page. Such materials may take the form of appendices, computer codes, animations, videos, etc., depending on the nature of the publication. Audio-visual materials may be used creatively to enhance the visual impact and accessibility of an article. (The ultimate example of this approach is to be found in the *Journal of Visualized Experiments* (JoVE), which publishes peer-reviewed scientific and educational videos of laboratory protocols and experimental techniques.)

12.4 Structure of an archival paper

Advice on how to prepare a manuscript for publication is readily available elsewhere. In the bibliography at the end, I have listed three articles that have appeared in leading scientific journals (Brand and Huiskes, 2001; Earnshaw, 2012; Whitesides, 2004). Many publishers provide online resources such as videos and webcasts to advise authors on matters ranging from research ethics, authorship, the preparation of a manuscript and its structure, to the use of appropriate language. Although manuscript formats do vary from discipline to discipline, scientific reporting usually follows the canonical structure: Introduction, Methods, Results and Discussion (IMRaD). Most papers include a Conclusions section in which the main conclusions of the work are summarised.

The introduction section should begin with a brief statement on the motivation for the study, and the specific research question(s) or issues that your study aims to address. If appropriate, these questions may be expressed in the form of one or more hypotheses to be tested. Covering too much material in one manuscript is frowned upon almost as much as too little. A manuscript that is too lengthy, contains too many figures, or is written in language that is too verbose, is likely to be rejected. Equally frowned upon is the over-interpretation of results, where the conclusions are not supported by the evidence. It is important, therefore, to consider the methods of statistical analysis to be employed at the planning stages, since this will determine the significance of any findings and the validity of any conclusions drawn. The information provided in the methodology section should be concise yet in sufficient detail to provide the reader with enough information to appreciate what was done and how. Clearly, one would expect a methods-based paper to focus on the technical aspects in greater detail, but this should not be at the expense of readability; additional information may be provided in an appendix, or online supplement as stated above.

The methods section should be followed by a section dedicated to the presentation of results, with each paragraph describing, in turn, the key findings or results pertinent to each research question. There should follow a separate discussion of the significance of those results in terms of the stated aims: do the results support or refute the hypotheses on which the study is based? Authors should explain the significance of their findings, making clear any major assumptions and/or limitations of the study approach or methods, and place their work into the wider context of the state-of-the-art, with reference to the work of others, as appropriate.

Only when the study is complete, and the entire paper has been drafted, should the abstract be finalised. Authors are often asked to provide a number of key words that may be used as search terms; remember that search engines will have access to the title and abstract as well, so choose your words with care. The importance of choosing appropriate keywords and terms cannot be overstated, since these will determine the ranking of your paper in the results of web searches; equally, the categories chosen at the time of submission may influence how your work is brought to the attention of other researchers. In this way, you can increase the chances of your work being cited.

12.5   Writing for publication: quality counts

Readability and concise writing are considered to be key attributes of a good research paper. Poor grammar and punctuation are a distraction for any reader, and native English speakers, especially. It goes without saying that authors should carefully proofread their manuscript before submission. Whilst many authors will readily admit that the feedback provided by reviewers helped to improve the quality of their manuscript, it is in the authors' interest to make the reviewers' job easier. It is unreasonable to expect reviewers, who are by definition active researchers themselves, to proofread your manuscript for basic errors in grammar and punctuation. A poorly written and constructed manuscript will raise questions in the mind of any reviewer that the same lack of care and attention has been applied to the work on which the manuscript is based. Apart from the damage to one's reputation, it wastes the precious time of all those involved in the peer review process. Authors for whom English is not their first language should consider the use of a professional English language copy-editing service to bring the work up to publication standard. Again, most publishers offer this service for a cost that is marginal compared to the cost of Open Access publication. (Such services should be distinguished from 'ghost-writing' services, the use of which is not acceptable, and no substitute for good scientific writing.) The manuscript should be carefully proofread in any case, ideally by a native English speaker. Your co-authors have a vested interest in proofreading the drafts of your manuscript, so do not be embarrassed to ask them in the first instance. A word of caution here: with the introduction of electronic submission, publishers routinely use plagiarism software (e.g., Turnitin®), which can detect similarities with previously published works, including those by the same research group. While a degree of overlap is inevitable, for technical papers especially, when a significant proportion of a manuscript is found to bear too close a resemblance to previously published works, authors risk being in violation of the publisher's policies on originality and plagiarism. This topic is covered at greater length in Chapter 13.

12.6   Peer review

When a manuscript arrives at the editorial office, it undergoes a preliminary check to determine that the authors have followed the submission instructions for the journal, and that it meets certain basic technical requirements (that the article does not exceed the word limit for the category under which it has been submitted, for example). These checks will include a basic English language check and the generation of a similarity report. The editorial staff will also check that any statements concerning any conflicts of interest/disclosure have been declared, and that prior approval by an institutional review body or ethics committee has been sought and obtained, as appropriate. Thereafter, the manuscript is assigned to an editor, who will assess the appropriateness of the work for the journal, its originality and timeliness. As many as half of all submitted manuscripts may be rejected outright at this stage if the editor has grounds for concern (poor quality, out of scope, etc.), otherwise the editor will proceed to invite potential reviewers from a database of several thousand reviewers on the editorial system. While invitations may be issued to reviewers nominated by the authors themselves, most will be selected based on their publication record and the key words provided by the reviewers when they first registered on the system. The database is continually updated, with reviewers being added or removed from the system on a daily basis.

The principles of peer review for journal publication are essentially unchanged since Henry Oldenburg, the founding editor and publisher of the first scientific journal in 1665 (*Philosophical Transactions of the Royal Society*; royalsociety.org), introduced the concepts of scientific priority and peer review as a prerequisite for publication. The peer-review process depends on willing reviewers to provide objective and timely reviews; however, it is not uncommon for reviewers to decline an invitation to review, and it may take several weeks before a sufficient number of reports have been received on which to make an initial decision. In my experience, about half of the manuscripts that are sent out for peer review are rejected at this stage. While, in my role as Editor of MEP, I am reluctant to make a decision on the basis of a report from a single reviewer, I believe there is a balance to be struck between providing feedback to authors and prolonging the peer-review process to a second, and sometimes third round of reviewer invitations. In such circumstances, I may be prepared to make a decision on the basis of a single report, normally when the recommendation is an unequivocal rejection. Manuscripts that are invited to undergo revision stand a far greater chance of being accepted, although one or more rounds of revision may be necessary before the manuscript reaches the standard required for publication. Whilst an invitation to revise a manuscript does not guarantee that it will be accepted, the majority of manuscripts in this category are published eventually.

Most journals accept fewer than one-third of the articles they receive. (For the *New England Journal of Medicine*, or *British Medical Journal*, for example, the rate of rejection is closer to 90%!) Authors who submit a manuscript for review will more likely than not be disappointed, so do not be too dismayed when this happens to you. When faced with this situation, it would be unwise simply to resubmit your work to the next journal on your list without taking the reviewers' comments into account since the same reviewer may be approached by another journal. This is not as unlikely as it may seem: reviewers are selected for their expertise, which is based more often than not on their publication record in the same or similar field. More than once has a reviewer complained to me that they had reviewed the same paper for another journal, yet the reviewer feedback had been ignored, or worse, the manuscript was identical to the version that was previously rejected. (I can think of examples from my own experience where the corresponding author did not bother to update the journal title and addressee in the accompanying cover letter!) Increasingly, publishers offer an 'article transfer service' whereby editors may recommend that a manuscript is passed to another journal in the publisher's stable, either before or after peer review. This generally happens when a manuscript is deemed to be more appropriate for that journal's audience and scope. Of course, authors are at liberty to withdraw their manuscript and submit elsewhere, though that does not guarantee that another editor will not approach the same reviewers in any subsequent review.

12.7   Conclusion

To be a successful academic, one must be able to communicate effectively in writing. This is a skill like any other, and takes time and a good deal of practice to master. Ask anyone who writes for a living! And whereas the adage 'Publish or Perish' may be as relevant to researchers today as in the past, it is no longer sufficient to get work into the public domain, since the impact of those research outputs, as measured by citations, is increasingly the benchmark against which a researcher's contributions are assessed. Thus, more than ever, it is in every researcher's interest to secure endorsement from their peers by ensuring that their papers reach the audience most likely to read, and cite, them.

References and bibliography

Brand, R.A. and Huiskes, R. (2001) Structural outline of an archival paper for the Journal of Biomechanics. *Journal of Biomechanics*, Vol. 34, 1371–1374.

Earnshaw, J.J. (2012) How to write a clinical paper for publication. *Surgery (Oxford)*, Vol. 30, Issue 9, 437–441.

HEFCE (2009) *Research Excellence Framework: Second Consultation on the Assessment and Funding of Research*. HEFCE 2009/38.

Hirsch, J.E. (2005) An index to quantify an individual's scientific research output. *Proceedings of the National Academy of Sciences of the United States of America*, Vol. 102, Issue 46, 16569–16572.

Schimel, J. (2012) *Writing Science: How to Write Papers That Get Cited and Proposals That Get Funded*. Oxford University Press.

Whitesides, G.M. (2004) Whitesides' Group: Writing a paper. *Advanced Materials*, Vol. 16, 1375–1377.

Support for authors

Elsevier Publishing Campus. Available at https //www.publishingcampus.elsevier.com

The peer-review process

Elsevier. Elsevier Reviewer Hub. Available at https://www.elsevier.com/reviewers

PLOS Medicine. Reviewer guidelines. Available at http://journals.plos.org/plos medicine/s/reviewer-guidelines

13 Publication ethics and academic integrity

Richard A. Black

13.1 Introduction[1]

I write this as a biomedical engineer with research experience spanning more than 25 years. In 2012, I was appointed Editor-in-Chief of *Medical Engineering & Physics*, one of three journals published by the Institute of Physics and Engineering in Medicine (IPEM). The scope of the Journal is broad, serving both the biomedical engineering and medical physics research communities, and reports on the latest developments in the field. With the support of the Editorial Board, Journal Manager and staff in the Editorial Office, the Journal processes around 700 original submissions each year, and publishes around 200 articles following peer review. The peer-review process is a truly multidisciplinary and international effort, involving authors, editors and reviewers from across the globe, the vast majority of whom selflessly participate in this vitally important work, which helps to ensure the quality of articles published, and that the highest standards in reporting are maintained.

It is a privilege to undertake this work on behalf of the Institute. The experience has not been wholly edifying, however. Sadly, it is not uncommon that we encounter problems on the part of authors and reviewers alike. Whereas the majority of articles report original work, regrettably a number show some evidence of redundancy or plagiarism. Authors are required to attest to the originality of a manuscript, and to declare that it has not been submitted for publication elsewhere; nevertheless, it is surprising how often concurrent or duplicate submission occurs. This practice is unacceptable, not least because it wastes the valuable time and resources of everyone involved in the peer-review process. With the introduction of electronic submission, publishers routinely submit manuscripts to search engines (e.g., iThenticate® – www.ithenticate.com) that can detect text plagiarised from published works. Unfortunately, software alone is unable to detect every form of copying, and the peer-review process relies on reviewers to raise any issues or concerns they may have about the work they undertake to review. When such cases are brought to the attention of the editor while a manuscript is under review, the review process may be terminated and the manuscript returned to the authors. In cases where the article has already been published, authors are invited to respond to any allegation before further steps are taken; thereafter, depending on the authors' response, editors may consider it necessary to publish a notice, or retraction. Fortunately, this is a rare occurrence, and I am not aware of any of my predecessors having had to resort to

---

[1] This introduction is based in part on an editorial published in 2014: *Medical Engineering & Physics*, Vol. 36, 3.

such sanctions in the pages of the journal for which I am editor.

Central to the peer-review process is the review by an expert of work submitted for publication, and that process depends on reviewers to provide objective and timely reviews if it is to function effectively. Although some efforts are being made to address the situation, it is fair to say that reviewers at present do not get the recognition they deserve for the work they do. Not surprisingly, therefore, invitations to review are more often than not declined or ignored – or, worse, accepted and then ignored, for reasons that may not be entirely honourable. While this may simply be due to pressure of work, it could also be seen as an attempt to gain advantage or simply to delay publication of a competitor's work.

I am aware, too, of several instances where authors have nominated former colleagues (a recent former co-author or academic supervisor, for example) to review their work. The use of false email addresses that redirect correspondence back to the authors is a more serious example of this trend. Some reviewers may choose to disclose their identity in their correspondence with authors, while others may attempt to garner citations to their own work by demanding that authors cite one or more of their previous works, and in doing so leave the authors in no doubt as to the reviewer's identity. Whether unintentional or otherwise, such practices are unacceptable, and undermine the integrity of the peer-review process. The obligations on the part of authors and reviewers alike are made clear in the author guidelines and on journal websites. These include, for work involving human subjects, the need to obtain prior approval by an institutional review body on matters of ethics, and informed consent from participants; declaring any conflict of interest, perceived or actual; and acknowledging source(s) of funding, etc.

The majority of scientific journals (and grant awarding bodies) operate a system of anonymous peer review, known as single-blind peer review, which protects the anonymity of the reviewers. Some adopt a double-blind review process in an attempt to conceal the authors' identity from the reviewers, but this is often difficult to achieve in practice. There are alternatives to anonymous peer review: open; open and published; and post-publication peer review, where comments are invited from the readership and published online. A more radical suggestion is to abandon peer review altogether in favour of usage statistics (downloads, citations, etc.). Arguably, the entire peer-review process is in need of an overhaul, but until such time as the alternative systems are shown to be more effective, and less open to abuse, anonymous peer review is likely to prevail for the foreseeable future.

13.2 Publish and be damned

Despite efforts by the scientific community to tackle misconduct in research, there have been several high-profile cases in recent years involving leading scientists and their publications. The prolific output of the physicist Jan Hendrik Schön of

Bell Laboratories between 2000 and 2002, for example, aroused suspicions in the scientific community when researchers elsewhere were unable to reproduce his findings. Closer examination revealed that some figures in different papers were similar, while others were based on data that was thought to be too 'clean' to represent real experimental variation. The committee that was convened to examine the evidence determined that Schön had fabricated data in at least 16 of 25 published papers, all of which were subsequently retracted. Other notable examples include the controversy surrounding the MMR (measles, mumps, and rubella) vaccine following publication in 1998 in *The Lancet* of a paper that reported an association between the incidence of colitis and autism in children who received the combined MMR vaccination. The paper was subsequently retracted when it transpired that the clinical trial data on which the paper was based were obtained unethically, and the lead author, Andrew Wakefield, had been funded by certain vested interests. Dr Wakefield was found guilty of professional misconduct and struck off the Medical Register. Nevertheless, the damage had been done: the misplaced disquiet of parents about the use of MMR led to a dramatic reduction in vaccination rates in the UK in the years that followed, giving rise in 2012 to the highest number of cases of measles in England and Wales in over two decades. The author of another paper published in *The Lancet* in 2006, Jon Sudbø, was found guilty of academic dishonesty when he was found to have fabricated the 900 patient records on which the paper was based. The Faculty of Medicine at the University of Oslo, where he was an associate professor, found that his 2001 doctoral thesis was likewise fraudulent. In all, 12 of Sudbø's papers and his PhD thesis were retracted. Another case brought international notoriety to the South Korean stem-cell scientist, Hwang Woo-suk, when a colleague claimed that Hwang had fabricated parts of a study into cloning of human embryonic stem cells. Several publications, including two in the journal *Science* in 2004-05, were later retracted. Dr Hwang was convicted of fraud and embezzlement. More recently still, is the case of thoracic surgeon Paolo Macchiarini, who has been under investigation by the Karolinska Institute in Stockholm for his role in experimental procedures to implant artificial tracheas in patients, following allegations of scientific misconduct relating to claims made in papers published by Macchiarini and his co-workers between 2011 and 2014.

The cases outlined above are unique, and not confined to any one culture, country, or region. Nevertheless, all of the individuals concerned were called upon to answer allegations that were raised after their work entered the public domain, and to appear before committees convened by the individual's employer, as well as professional bodies and, ultimately, courts of law.

13.3   Publication ethics: possible sanctions

As one would expect, editors take allegations of misconduct very seriously. Most leading publishers are members of the Committee on Publication Ethics (COPE; www.publicationethics.org), which was established to offer support and guidance to journal editors, but equally to authors and reviewers. There are several case studies on the COPE website for readers to examine, along with their outcomes. The options available to editors faced with a *prima facie* case of plagiarism or other forms of misconduct range from the publication of an editorial, notice, correction or *corrigendum*. It is left to the discretion of the editors and editorial boards to apply more punitive sanctions, such as barring further submissions from author or author's group. In most cases, however, the publication of a notice to correct the scientific record is sufficient to draw attention to the matter. It is left to readers to determine the extent to which the authors' reputations may have been tarnished as a result.

According to COPE, authorship is a common concern, with many of the case studies mentioned above featuring headings such as 'authorship', 'changes in authorship', or 'disputed authorship'. The COPE discussion document 'What Constitutes Authorship' (COPE, 2014) offers a useful guide on this difficult issue. In some cultures, researchers may be inclined to extend authorship to colleagues or collaborators when an acknowledgement may be more appropriate – for technical assistance or assistance with the writing or proofreading a manuscript, for example. Guidelines published by the International Committee of Medical Journal Editors (ICMJE; www.icmje.org) stipulate that authorship demands the intellectual input of all authors to the conception, design and execution of a study, as well as the analysis and interpretation of results (ICMJE, 2015). This requirement extends to the drafting of any publications arising from that work, and input to any revisions that may be required before its publication. All authors are expected to be held accountable for their individual contributions and vouch for its accuracy and integrity, and to work to ensure that any disputes are investigated and resolved.

While it is generally accepted that authorship must be earned, what determines authorship is sometimes harder to pin down. Drawing on my own experience as an editor, I have found it necessary to invite more than one corresponding author to explain why a name has been added to the list of authors in revision for no apparent reason. In such cases, it is incumbent on editors to write to all of the authors concerned in order to establish their consent, in writing, to those changes. I am aware also of instances where, on completion of the peer review process, the order of names appearing on a manuscript is altered, often at the last minute. While this may be due to oversight, peer pressure undoubtedly plays a role. Some institutions, for example, place great store when considering cases for promotion on the number of papers where an author is named either first or last in the running order, with less credit being given to authors whose names appear in between, regardless of their contribution, and despite the ICMJE requirements mentioned above. An unwritten rule is that the researchers (RAs, PhDs) are named in descending order of contribution followed by the remaining authors in ascending order of contribution usually with the person leading the work (i.e. principal supervisor or principal investigator) as last-named author. It is possible, however, for the contribution of the first two authors to be recognised and granted equal status as appropriate. The role of corresponding author, too, is sometimes seen as an indicator of status. As the name suggests, the corresponding author is responsible for submitting the paper to the journal, and keeping all authors informed of its progress throughout the peer-review process, responding to the reviewers' comments, and seeking their input when revising the manuscript until a decision is reached. While practices do vary from discipline to discipline, it is in all of the authors' interest to agree authorship from the outset if only to avoid any misunderstandings or ill feeling later on. Inclusion of an author who has made no contribution to the paper, in the accepted manner, is known as 'gift authorship', and is considered unethical.

13.4   Values, standards and practices: The Nolan Principles

These principles are based on a Code of Practice developed by a Parliamentary Committee on Standards in Public Life set up by the UK Government in 1994, chaired by The Rt Hon The Lord Nolan. The government of the day was embroiled in a series of political scandals, characterised in the UK press by the term 'sleaze', and the committee was charged with drawing up a set of standards to which elected representatives, government officials, essentially all those in the public eye, are expected to subscribe. The 'seven principles of public life' identified by the Nolan Committee are: Selflessness; Integrity; Objectivity; Accountability; Openness; Honesty; Leadership.

The current chair of the committee is Lord Bew, one of whose predecessors, Sir Christopher Kelly KCB, was tasked with investigating irregularities in MPs' claims for expenses. To quote from the Committee's definition of selflessness: 'Holders of public office should act solely in terms of the public interest. They should not do so in order to gain financial benefits for themselves, their family or their friends.' Clearly, the members of the committee that drafted these guidelines could never have imagined that this might include 'feathered' friends! (A reference to ducks and duck houses purchased by a member of the House of Commons on expenses...)

13.5   Codes of practice for research

In response to several high-profile scandals involving leading public figures, scientists and clinicians outlined above, most universities and professional bodies have instituted clear policies on such matters, and procedures that are to be followed in cases of suspected professional dishonesty or misconduct. Government agencies, too, have published a series of articles and publications on the principles of good scientific practice: in the US, the National Academies Press published *On Being a Scientist*, now in its third edition; and in the UK, *Promoting Good Practice and Preventing Misconduct* is published by the UK Research Integrity Office.

13.5.1 IPEM's code of conduct and disciplinary procedures

The expectations placed on medical physicists and engineers are not limited to matters of publication ethics. Members are expected to adhere to the same principles that apply to all those in the public eye, all the more so in cases where the well-being and treatment of patients is concerned. It is incumbent on professional bodies and learned societies such as IPEM, therefore, to ensure that its members practise with the highest standards of integrity, in keeping with best practices across the world. Such good scientific practices require researchers to:

- maintain professional standards;
- secure and store primary data;
- document results;
- interpret results objectively;
- attribute honestly the contribution of others; and
- be aware of any relevant ethical considerations.

The regulatory bodies for the UK engineering and scientific professions, the Engineering Council and Science Council, operate under Royal Charter to ensure professional competence and standards of ethics of practitioners in those professions. IPEM is licensed to assess its members for registration on behalf of those organisations, which maintain national registers of those who can demonstrate that they meet the internationally recognised standards expected of medical physicists and engineers. All members are required to comply with the Institute's *Code of Professional and Ethical Conduct* (IPEM, 2015), and may be subject to the Institute's Disciplinary Procedure when in breach of this Code. The Code has 24 clauses, relating to their interaction with patients, colleagues and the public at large, which members of the Institute are expected to uphold. The Code is organised under the following headings:

- Patients, clients and users;
- Rigour, responsibility, honesty and integrity;
- Respect for life, the law, colleagues and the public good;
- Personal and professional life; and
- Responsible communication: listening and informing.

Those members whose work may have a bearing on the treatment of patients are referred also to the Institute's *Guidelines for Good Practice: Working with Patients* (IPEM, 2016).

These documents make clear the values, standards and practices expected of those engaged in research, which include all those who work in publically funded institutions such as the NHS and universities. Understanding these practices, and the standards that underpin them, is key to a successful career in research.

References and Bibliography

COPE (2014) *What Constitutes Authorship? COPE Discussion Document.* Committee on Publication Ethics. Available at http://publicationethics.org/files/Authorship_ DiscussionDocument.pdf

ICMJE (2015) *Recommendations for the Conduct, Reporting, Editing, and Publication of Scholarly Work in Medical Journals.* International Committee of Medical Journal Editors. Available at http://www.icmje.org/recommendations/

IPEM (2015) *Code of Professional and Ethical Conduct.* Institute of Physics and Engineering in Medicine, Document 0217, 6th issue, January 2015. Available at www.ipem.ac.uk

IPEM (2016) *Guidelines for Good Practice: Working with Patients.* Institute of Physics and Engineering in Medicine. Available at www.ipem.ac.uk

Research ethics

American Physical Society: http://www.aps.org/programs/education/ethics/resources.cfm

European Association for Chemical and Molecular Sciences. *Ethical Guidelines for Publication in Journals and Reviews*: http://www.euchems.eu/wp-content/uploads/Ethicalguidelines_tcm23-54057.pdf

Royal Society of Chemistry. *Code of Conduct & Conflicts of Interest*: http://www.rsc.org/journals-books-databases/journal-authors-reviewers/author-responsibilities/#code-of-conduct

The National Academies Press. *On Being a Scientist. A Guide to Responsible Conduct in Research.* 3rd edition. The National Academies Press, Washington, DC. Available at https://www.nap.edu/catalog/12192/

UK Research Integrity Office. *UKRIO Code of Practice for Research: Promoting Good Practice and Preventing Misconduct*. UK Research Integrity Office, September 2009. Available at http://ukrio.org/publications/code-of-practice-for-research/

US Department of Health & Human Services, The Office of Research Integrity: http://ori.hhs.gov/

14 The growth of the 'impact' agenda in British universities and research in the UK

Alicia El Haj

14.1 Background

In 2012, HEFCE (Higher Education Funding Council for England) published the new *Assessment Framework and Guidance on Submissions* in preparation for the 2014 Research Excellence Framework (REF) exercise for research in UK universities (HEFCE, 2012). A major change from previous research assessment exercises was the addition of a category termed 'Impact' which would frame a shift in the research emphasis within universities towards a greater role in influencing or benefiting the surrounding national and international communities. The impact would be in the form of the growth of businesses, the economy, government, civil society and the development of public and policy debates. The shift in assessment crossed all disciplines within the universities, not only the more traditional engineering and medical subjects well versed in translational research but also social science and humanity subjects such as languages and philosophy. The initiative established that 20% of government research support to universities, or QR (quality-related research) funding, would be based on a scoring of how far departments and research units have achieved external impacts.

Interestingly, the expectation of impact did not extend to each member of academic staff submitted. It was assumed that impact could be delivered by groups or collaborative teams of academics with one example of impact required per 10 staff submitted. This equates to one impact per approximately 40 publications from a unit, which in some institutions would not be a high bar to cross but in others would be potentially challenging. In total, 6679 case studies were submitted from 154 institutions, ranging in number from 2 to 260 per institution. The tool for assessment of the significance of impact was the 'impact case studies' pro-forma, requesting information going forward to assessment. The 2014 HEFCE sub-panels assessed the 'reach and significance' of impacts that were underpinned by excellent research within the grouping, as well as the group's approach to enabling impact from its research. A key element of the outcome of this new approach is to align substantial government money to departments which have done well in applied economic, government and public facing activities. Appendix 14.1 provides further details of 'impact', from the REF 2014 submission guidelines.

14.2   The impact template

In order to achieve consistency in the information which was submitted, HEFCE created a template for impact case studies. Each impact case study was graded from 0 to 4 star in quality based on metrics. The impact template asked for key elements which included a summary of the impact, a detailed description, and the context of the significance. The request also included information on the underpinning research in the form of a list of publications leading to the impact. These publications would be graded according to the traditional 1 to 4 star system applied to publication outputs in the REF. If research was not found to be above 2 star quality, then the case study was not assessed or scored. The assumption was that excellent research will lead to excellent impact, which was widely debated by some institutions.

In addition, the details of grant funding over the past 10 years which led to the impact were requested. This has provided an excellent resource of data to track funding from research councils in the UK and Europe through the translation and commercialisation process. Finally, and perhaps less commonly requested in previous submissions for research data, was a section listing 'sources to corroborate' the impact (up to 10). These sources might be contacted to check on the accuracy of the data submitted and so were an essential aspect. Contacts could include key personnel from industry, government, clinicians, etc.

Appendix 14.2a shows the web link for a submitted case study from the Institute of Science and Medicine at Keele University which is an example of innovative cell therapy which has had a clinical impact through translational research within the group at Keele. The key element of this example is the long development time through successful grant funding. The research was a collaborative programme between surgeons, clinical scientists and bioengineers which took over 10 years to move from basic studies to clinical application with interactions with key UK industries and start-ups resulting from the research. In this example, multiple clinical centres had adopted the approach and over the past 10 years, hundreds of patients have benefited from the therapy. This case study illustrates a number of key points which include the importance of multidisciplinary research teams and the significant funding required to reach impact in the healthcare system. Appendix 14.2b shows the web link for a case study from King's College London which demonstrates impact in the area of medical imaging showing how the multi-modality imaging approach of MRI and PET was researched and adopted as a result of the work conducted at King's. This is an excellent example of high quality research moving through to preclinical and clinical trials before adoption as a routine technique within multiple hospitals in the UK.

14.3   Assessment – reach and significance

The impact case studies were assessed on 'reach' and 'significance', two terms which may be interpreted in many ways. Essentially, the term 'reach' covered the breadth of the impact: the mobile phone, which has reached billions of people around the world would be an extreme example. Potentially, novel imaging modalities, which had been adopted across the UK and the western healthcare systems, would also be an excellent example as would potential therapies or medical devices adopted routinely into medical environments. Of course, the majority of medical translational research takes significant periods of time to reach the clinic, hence for a product to have reached the stage where a small number of patients can be treated in a safety trial is considered extremely successful. With this in mind, the term 'reach' was tailored towards disciplines with examples from basic engineering having a greater expectation for a 20-year development window of 'reach' than in a clinical context where it may be even longer. The term 'significance' was an assessment of the level of advancement of the impact. In the case of cell therapies outlined in the link to a case study in Appendix 14.2a, this type of treatment is considered a 'step change' to existing practice enabling new approaches and treatments for diseases with few existing therapies available currently. Imaging modalities enabling a new form of visualisation at perhaps increased resolution within the body would also be examples of work with major significance (case study, Appendix 14.2b). New medical implants or devices are further examples of this type of significant impact.

The activity should go beyond 'business as usual' engagement or involvement, for example, if there was active involvement of end-users and/or the public, the activity informed the focus of the research or created widespread interest, was particularly innovative, or created a legacy of resources (Figure 14.1) (King's College London, 2015).

14.4   Outcomes in healthcare

In the 2014 REF assessment, 154 universities were assessed and the metrics from the analysis have been published in an online database (http://results.ref.ac.uk/; http://www.hefce.ac.uk/pubs/rereports/Year/2016/refimpact/Title,108841,en.html).

In the recent report by King's College London on the 2014 impact case studies (King's College London, 2015), the group researched through the impact database to assess 'what the impact and value of research has been on clinical practice and health gain'. Of the 6679 case studies submitted, 6% of the total (416) were labelled as healthcare services and 5% (326) as clinical guidance. The report identified that there are multiple ways for biomedical engineering and biomedical research to translate from bench to bedside. An important way is to inform clinical guidelines, for example, the National Institute for Health and Care Excellence (NICE) guidelines which inform clinical practice. NICE is a non-departmental public body

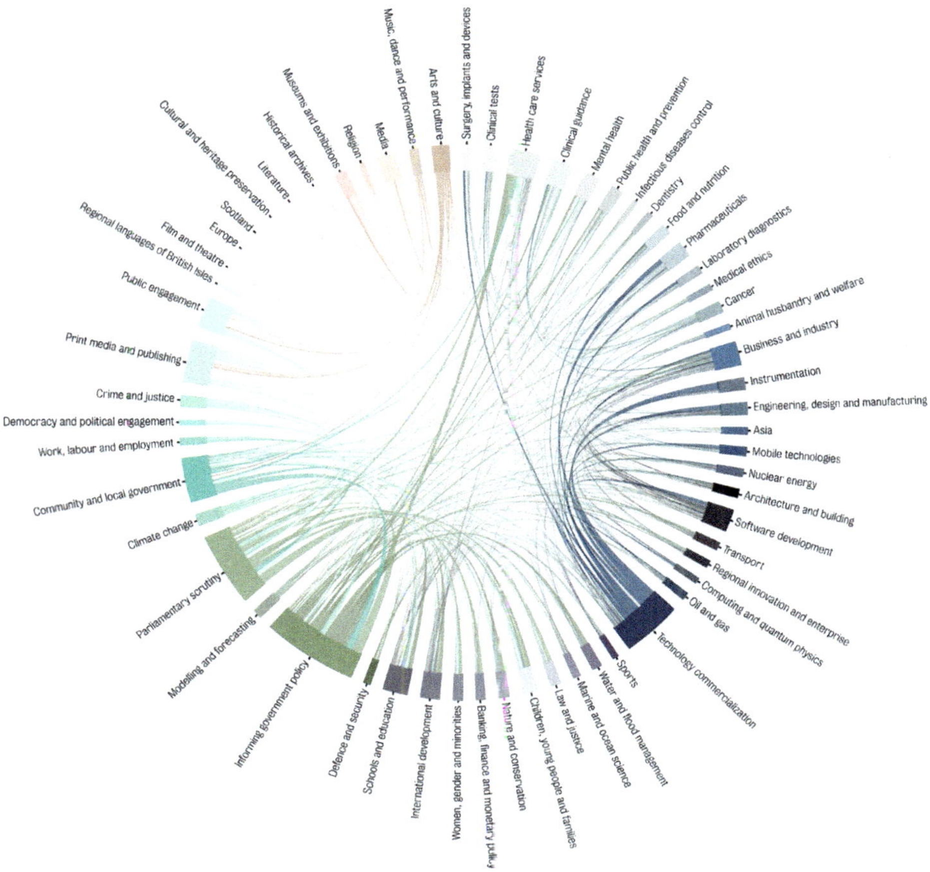

Figure 14.1 Chord diagram of the size and relationship of impact across the breadth of disciplines within universities in the UK – the colours denote the intensity of the impact in each discipline with darker colours indicating high intensity and lighter colours indicating a lower intensity. Reproduced with permission from: King's College London and Digital Science (2015) The Nature, Scale and Beneficiaries of Research Impact: An Initial Analysis of Research Excellence Framework (REF) 2014 Impact Case Studies. Research Report 2015/01

and is responsible for developing guidance and quality standards in health and social care. Another way is to measure the impact through quantitative metrics used in healthcare by economists, termed QALYS (Quality Adjusted Life Years). The report searched through the database and assessed those that had used this quantitative measure estimating a figure of £2 billion in the impact period 2008–2012.

Figure 14.2 The global reach of impact from British higher education institutions (HEIs) – numbers indicate the times each country was cited in an impact case study as having impact. Reproduced with permission from: King's College London and Digital Science (2015) The Nature, Scale and Beneficiaries of Research Impact: An Initial Analysis of Research Excellence Framework (REF) 2014 Impact Case Studies. Research Report 2015/01

Figure 14.2 indicates the number of impact case studies which demonstrated reach into different countries across the world. USA, UK and Australia had the greatest degree of reach and impact as shown by the dark blue colour.

14.5 Challenges

The REF definition of 'impact' often requires a broader remit of relevance, for example, in the case of the patient population. In clinical studies, the involvement of patients is termed patient and public involvement/engagement (PPI). PPI groups are often part of a wider effort to define protocols for translational clinical trials. What is not clear is the level of 'impact' of PPI on the 'outcomes' as opposed to 'protocol development' in research. Indeed, it may be difficult to demonstrate a causal relationship between the involvement of PPI in a process and the impact (Caress, 2016).

Another challenge is obtaining the relevant 'evidence'. Examples of relevant sources may include demonstrable changes in policy or practice, media reports or testimonials from relevant stakeholders. Many researchers found it difficult to provide sound evidence of impact, as this has often not been adequately captured over the past period of 20 years.

In the recent Stern review published on 28 July 2016, the committee identified that linking research outputs such as publications to impact could be very confining for institutions. In many cases, universities can have significant impact but based on a broader context than a 2 star publication (King's College London, 2015).

In addition, the cost of gathering the impact studies for universities has been calculated at £55 million. There are discussions as to whether this money is well spent in gathering the data or if there may be less costly ways to measure and assess impact (King's College London, 2015).

Another example of a challenge with writing these impact studies was tracking the impact over a 10- to 20-year window. This could include searching information from industry on sales and distribution figures which could often be commercially sensitive data or tracking clinician users across a wide breadth of British universities. Also, sources for corroboration could often change jobs or positions with universities.

Finally, a criticism that is sometimes levelled at the impact agenda is that it potentially discourages publication of negative results. Negative results have a very important place in science, indicating to other researchers approaches that are not fruitful and so saving wasted effort and expense. They are particularly valuable in medical research, where knowing that a particular treatment does not improve patient survival is of obvious benefit to future patients who might otherwise be enrolled in trials of drugs that have already been shown not to work in previous unpublished research. The AllTrials campaign (www.alltrials.net/), which is supported by IPEM, is calling for the results of all clinical trials to be published for this reason. But clearly, negative results are much less likely to score highly in terms of impact than positive ones that go on to significantly change an area of medical practice and to improve patient outcomes. Currently, trials with negative results are twice as likely to remain unpublished as those with positive results, and the resulting bias in published results can be clearly seen using what is known as a funnel plot (Egger *et al.*, 1997). There are concerns that the impact agenda promotes this anomaly.

14.6   The university impact environment

The assessment exercise included requests for details of the academic impact environment (HEFCE, 2012). Information on the support for and approaches to creating influence and impact outward from the university was assessed. In particular, the case studies were placed in the context of the returning unit, the unit's approach to impact during the period 2008–2013 was monitored, and the overarching university strategy and plans for supporting impact were assessed. Finally, the relationship between the department's or university's approach to impact and the submitted case studies was defined in the submissions. This was often clear, with a forward looking and level vision. This was especially important in the case of interdisciplinary research and translational work to define routes which span across multiple forums and units.

14.7 Future recommendations and the 2021 REF assessment

At the end of July 2016, the Stern review published the outcome of a reflective committee which looked at the REF process in 2014 and asked how it could be improved (Institute of Historical Research, 2016). The outcome is general confirmation that the REF is a worthwhile and strong exercise. It is unlikely that the broad structure will change. It is clear that the impact agenda will only grow as this type of translational activity from universities is recognised and rewarded. A number of recommendations have been made for changes to how impact is monitored in these quality exercises. The report recommended a decoupling of the link between research and impact to enable a broader base of evidence and supporting work to be considered. This restriction during 2014 was felt not to enable the breadth of the impact agenda across UK universities to be captured. In addition, the committee recommended that case studies should be presented at an institutional level with a specialist cross-disciplinary panel being formed to assess them. Finally, the committee stated that potentially using existing case studies but broadening them would be possible in 2021.

The outcome of this REF 2014 review is still being discussed with a new timetable aiming towards the 2021 exercise now in place (Department for Business, Energy & Industrial Strategy, 2016). In a long development pathway as with healthcare, this would indicate that there may be another 7 years to really demonstrate the impact of UK universities on NICE and QALY statistics for new interventions and therapies.

References

Caress, A.-L. (2016) *Public Involvement and the REF Impact: Squaring the Circle.* Available at http://www.invo.org.uk/public-involvement-and-the-ref-impact-agenda-squaring-the-circle/

Department for Business, Energy & Industrial Strategy (2016) *Building on Success and Learning from Experience: An Independent Review of the Research Excellence Framework.* Available at https://www.gov.uk/government/publications/research-excellence-framework-review

Egger, M., Davey Smith, G., Schneider, M. and Minder, C. (1997) Bias in meta-analysis detected by a simple, graphical test. *British Medical Journal*, Vol. 315, 629–634.

HEFCE (2012) *Assessment Framework and Guidance on Submissions.* REF 02.2011. Available at http://www.ref.ac.uk/about/guidance

Institute of Historical Research (2016) *Stern Review: An Initial Bibliography.* Available at http://blog.history.ac.uk/2016/07/stern-review-an-initial-bibliography/

King's College London (2015) *The Nature, Scale and Beneficiaries of Research Impact: An Initial Analysis of Research Excellence Framework (REF) 2014 Impact Case Studies.* Research Report 2015/01. King's College London and Digital Science. March 2015. Prepared for HEFCE, HEFCW, SFC, DELNI, RCUK and the Wellcome Trust. Available from http://www.kcl.ac.uk/sspp/policy-institute/publications/Analysis-of-REF-impact.pdf

Appendix 14.1. Definition of impact for the REF – excerpt from the REF 02.2011 'Assessment Framework and Guidance on Submissions' (HEFCE, 2012)

For the purposes of the REF, impact is defined as an effect on, change or benefit to the economy, society, culture, public policy or services, health, the environment or quality of life, beyond academia.

Impact includes, but is not limited to, an effect on, change or benefit to:

- the activity, attitude, awareness, behaviour, capacity, opportunity, performance, policy, practice, process or understanding
- of an audience, beneficiary, community, constituency, organisation or individuals
- in any geographic location whether locally, regionally, nationally or internationally.

Impact includes the reduction or prevention of harm, risk, cost or other negative effects.

For the purposes of the impact element of the REF:

a. Impacts on research or the advancement of academic knowledge within the higher education sector (whether in the UK or internationally) are excluded. (The submitted unit's contribution to academic research and knowledge is assessed within the 'outputs' and 'environment' elements of REF.)

b. Impacts on students, teaching or other activities within the submitting HEI (Higher Education Institution) are excluded.

c. Other impacts within the higher education sector, including on teaching or students, are included where they extend significantly beyond the submitting HEI.

Appendix 14.2. Examples of REF 2014 submitted impact case studies

a. General Engineering B15 – Biological and Cell Therapies in Orthopaedics – Autologous Chondrocyte and Mesenchymal Stem Cell Implantation. Available at http://impact.ref.ac.uk/CaseStudies/CaseStudy.aspx?Id=6264

b. General Engineering B15 – Simultaneous PET & MRI. Available at http://impact.ref.ac.uk/CaseStudies/CaseStudy.aspx?Id=41229

Section 3

Research funding and translation

15 Research funding

Keith M. McCormack

If money go before, all ways do lie open

W. Shakespeare, *The Merry Wives of Windsor*, 1602, Act 2, Scene 2

15.1 Introduction

It is an unpalatable truth that, just as in that other less important – so-called 'real' – world, it's money that makes research go round. This chapter looks at the significance of funding to the modern researcher's way of life, considers the challenges from several perspectives, and suggests some strategies for survival, even for success.

As an engineering or physics professional in medicine, your training will have equipped you for either (or maybe both...) a career in clinical science and engineering, probably with leanings towards technical development, or – if you chose an academic career – a life already dedicated to investigation. In both cases, your steps into research will follow a path that to some extent will bounce you, pinball-like, between islands of funding that will sometimes make it seem very little like the well-directed straight-line journey for which you might have hoped. Yet there are some sensible steps, some logical approaches, and some useful techniques that can help to smooth out the bends.

15.1.1 The academic dream

If you followed your academic instincts, you probably did a PhD and went on to be a post-doc, you've already published successfully, obtained a little teaching experience, developed a conference profile, and tried a bit of project management. Your richly-worded CV and well-practised interview technique have secured you your first academic post, and now it's time to start building your independent research profile, establishing your first team and grounding your career in order to satisfy that urge to discover. Hideously, across all of engineering and beyond, this means you need funding. Do you scrabble madly for anything anyone will give you, or is it time for a strategy?

15.1.2 The clinical reality

If you opted instead to apply your skills towards patient care, your job is perhaps already satisfying, and in any case leaves you little time for research activities. But this would be to miss the important opportunities that almost certainly exist all around you, for seeking deeper truths and developing novel solutions to challenges that ultimately can improve patients' lives. Structuring a research strategy to achieve this could be the most satisfying activity of your career.

15.2   Your area of research

Whether academic or NHS, when it comes to your plans for research, ideally you've identified your general territory of enthusiasm and now you're looking for a niche area in which to make a mark and develop a research theme. You're combining your academic leanings with those of your department, and you're looking for the mix of overlap and interest that will satisfy everyone. We live in a time both of new approaches – to data, to modelling, to complexity, to engineering itself – and of the continuing dissection of topics into a more granular assortment of subcategories, as it becomes increasingly difficult for any one person to embrace the totality of a subject. Arguably, this is ideal for the novice researcher, as by combining specialisation with novelty of technique, there can arise a ready-made opportunity for thematic development.

It is often considered futile to plan much beyond even the end of the year, as the pace of change makes a mockery of such ambition. Whilst this might just be true of the detail, it is nevertheless instructive to draft a framework of progress spanning perhaps 5 years or more, in which you establish a timetable for the steps that need to be taken if your research career is to progress, and your funding targets are to become increasingly ambitious (Table 15.1).

Table 15.1 Possible progression of clinical and research targets

Clinical stage	Clinical research targets	Academic stage	Academic research targets
Specialist training	Critical thinking and other skills First funding (travel, conferences)	Doctorate	Critical thinking and other skills Teaching First funding (travel, conferences)
First professional post	A little teaching? First publications	Post-doctorate	Teaching Publications
First staffed research project	Research niche developing	Lectureship	Research niche Strategic PhD project activities
Multiple researchers	Time for an academic career?	Senior lectureship	New course development

15.3 Understanding different perspectives

To provide a context for the consideration of funding possibilities, it may be useful to review the differing outlooks, ambitions and expectations of the various parties to any research activity. And although it is possible to consider each research project in isolation, it is probably more insightful to consider a bigger picture, a grander design into which your various schemes can be fitted. The following ratings are meant as a rough guide; you will need to make your own evaluation of individuals within each organisation as the key to collaboration is that you get on with someone and you trust them.

15.3.1 Your hospital

Your NHS employer is pretty firm: you have a clear set of clinical responsibilities that must be met, come what may. Yet you're working in an environment where to stand still is to fall behind. The organisation may move slowly, but it's in one direction – toward better patient care. They want to increase their portfolio funding (NIHR financial support for research), but do not have the resources to support you and expect you to be self-sufficient – and to sacrifice some free time.

Motivation: Success. Demeanour: They talk well. Support: Look hard for it. Trustworthiness: 7/10

15.3.2 Your university

Your academic employer has probably given you a mixed message: you have a clear set of teaching responsibilities that must be met, come what may. Yet you're working in a research-driven environment, and you'd better develop the research strategy that your department's future reputation depends on, and which is the only real reason for your existence. If you do not embrace the research ethos, you'll have a troubled existence.

Motivation: Success. Demeanour: Wolf as sheep. Support: Well-meaning, but soft. Trustworthiness: 8/10

15.3.3 Your colleagues

At best they're too busy to help you, at worst they want you to fail (you're competing for the same pot). But it's worth striving for that enduring partnership that means you always have someone to bounce your ideas against, who'll tell you when you're wrong. And it's rare for one individual to have all the skills needed to write a really strong, complex proposal. Perhaps the truly successful lone researcher must have multiple personalities?

Motivation: Success. Demeanour: Wolf as sheep. Support: Well-meaning, but soft. Trustworthiness: 8/10

15.3.4 Your funders

A mixed bag: funders need to be associated with success – even the European Union (EU) keenly seeks favourable reporting of results; but thankfully for patients, funders' efforts go to selecting winners rather than talking-up their failures, though some are not above a bit of spin. The downside is conservative assessment by unimaginative peers who favour teams with a track record. Getting to know the individuals in smaller charities is probably worthwhile, as they can give insights into their preferences. Despite the widespread obsequious behaviour often witnessed towards the officers of the larger funders, it serves no purpose, as their absolute commitment to transparency and fairness means that all decisions rest with their independent experts. Don't lower yourself.

Motivation: Success. Demeanour: Proper. Support: 'Read our rules'. Trustworthiness: 9/10

15.3.5 You (your career – and your calling?)

No-one is motivated to look out for you quite like… you. Your research career and your clinical or teaching activities will inevitably come into conflict. Your career is central, but you're reading this because research is important to you – try not to do it an injustice. The successful researcher almost certainly at some moments must resort to those real-world techniques of persuasion, temptation, over-enthusiasm, and exaggeration – sometimes without realising it – to bring financial and, it is hoped, research success. Try not to let this depress you. It's important to understand your own motivations (Curiosity? Anti-boredom? Status? Reward?), and many researchers find these change over time as they amass experiences – the good and the bad.

Motivation: Solid. Demeanour: Keen, busy. Support: Reliable. Trustworthiness: Perfect

15.4 Necessary skills

Winning funding for your research requires more than just a good idea, you need to develop a particular set of skills to break through the sometimes hidden barriers to success.

15.4.1 Critical thinking

Your proposal looks good to you – of course it does – but the reviewers think more critically. They must choose the best possible destinations for limited funding, their criteria are strict and their scoring is remorseless. You simply must reconsider everything you write, looking through the sceptical eyes of the reviewer who has seen it all before. Crucially, your central idea must unambiguously and definitively tackle the need that you have identified, and there must be no possibility that an alternative approach would yield superior results. You are permitted to fail, but you are not permitted to advance a substandard idea. Developing critical thinking is a fundamental requirement and to begin with, you may need your own, tame, critic to play devil's advocate. Consider becoming a reviewer yourself.

15.4.2 Writing ability

Without doubt the single most important skill to acquire is effective, persuasive writing. This rarely comes automatically, rather it is built on a set of techniques that can be learned. Most important is the ability to generate a sense of narrative that pulls the reader through, without letting up, told with a sense of authority.

- A gripping introduction – reviewers are (mostly) human.

- Sound evidence for your capability in the field.

- A logical approach.

- A clear description of the problem (the opportunity...).

- Excitement at the well-constructed way in which your planned work will take advantage of the opportunity (OK, will solve the problem).

- The substantial benefit (however measured – there are many methods) that will follow.

- The publication opportunities.

- The impact across multiple domains.

- The next steps, demonstrating an enduring area for rich, continuing development.

15.4.3 Learning to write: resources

If the list above seems daunting, help is available and you can learn to write well. The best way is to use your institution's services to obtain copies of successful proposals to the funder you have chosen and mimic the approach. Chapter 6 in this book provides advice for the novice writer. Beyond that, there are some valuable online resources.

- Every university publishes its own advice; many of these are helpful. Take a look.

- Basics: http://research.microsoft.com/en-us/um/people/simonpj/papers/Proposal.html

- Detail: http://www.imperial.ac.uk/research-and-innovation/research-office/preparing-a-proposal/

15.4.4   Networking, consortia, partners

Many complex grants require multiple applicants under the leadership of a coordinator. The UK Research Councils accept applications from university and NHS researchers, so be prepared to collaborate. Your first (non-travel) research grant could be a collaboration – often the most successful arrangement is NHS/university pairing. The key point is – be prepared to work in collaboration. You learn more than just science.

15.4.5   A note on interview technique

If you are selected for interview, you will receive much advice. Here is a surprising item: **Don't answer the question.** That is to say, do not leap to answer the question immediately. The best approach is probably to see every question as an opportunity to provide additional information that you could not fit into the proposal. The reviewer asks: 'You chose to do this; why?' 'Ah yes', you answer, 'we weighed-up multiple factors when reaching this decision; firstly, there was...' and you can now show how far-seeing and insightful you are, how skilled at selecting from a set of complex choices, and how well-suited to taking on a programme of research. Oh, and do not read out all the words on your slides.

15.5   Funder categories, characteristics, proposal types

You will seek to match your research challenge to the most promising funder. It follows that the characteristics of funders and their instruments need to be understood. Here is a brief introduction.

15.5.1   Small charities

The Association of Medical Research Charities lists 133 members (see Appendix 15.1), mostly small and highly focused. Make friends with the charities closest to your calling, and understand their motivations. Typically, they are very driven by their patient groups, and are seeking practical solutions to patient problems. Although not averse to longer-term proposals, by definition, they will not be able to fund large-scale activities, but will enjoy being associated with researchers who go on to greater things.

Proposal structures may be flexible, but calls tend to be infrequent and every relevant research group will be competing, every time. Nevertheless, a very useful source of, typically, early-career funding.

15.5.2   Large charities

The major medical research charities, of which there are perhaps ten, are powerful and important players in shaping the national agenda and focus of investment, development and progress. They are occasionally vulnerable to social whim, but they develop short/medium/long-term strategies, consult widely, value public opinion and publish their intentions regularly in thoughtful, well-reasoned plans.

Proposal structures are comparatively rigid, and it's important to understand the range of projects that are of greatest interest. Proximity to results is less important, but track record is very relevant.

15.5.3 Research Councils

These are now accessible by universities and research-active NHS Trusts, and research collaborations are possible. With obvious foci (from their names), they reluctantly collaborate with each other (important for crossover engineering/medical projects) but (in 2016) were in turmoil over their structural rearrangement and seemingly oblivious to the shift in national patterns of applications that will follow the UK's exit from the EU. RCs offer many opportunities for funded career development.

Proposals are highly structured, formulaic, but short. It can be difficult to tell a good story in eight pages.

15.5.4 European Commission

Grants funded by the EC are prestigious, independently reviewed, but driven by comprehensive adherence to every detail of the call, and requiring attractive scientific vision, demonstrable management and leadership, adequacy of the consortium, etc. The EC is sometimes criticised because the proposals must be so well structured and the outcome so predictable that there is perhaps no room for true research. Many funding instruments exist but, in the eyes of universities, EC Research and Innovation Actions are probably the pinnacle of success. Multi-partner international consortia are mandatory. Following the 2016 referendum decision for the UK to leave the EU, the future of EU funding to UK institutions is uncertain.

Proposals are challenging to create, typically running to well over 100 pages, and it is usual for the coordinating partner to accept the majority burden of text composition. Success rates are very low.

15.5.5 Industry

Diagnostic and therapeutic device companies, prosthetics manufacturers, clinical decision-support systems, hospital equipment makers... the list is long, and the relationships depend on mutual respect and benefit. Cultivating a long-term relationship with the industrial operators in your technology area is almost essential if your work is to benefit patients, but it's a complicated process and the road has many hurdles. Start simple, build trust, share experiences, and – most of all – help profits.

Proposals are completely dependent on the nature of the relationship, the work to be done, and the licensing possibilities. Industry will often support PhD projects (perhaps via an EPSRC CASE award). Build the relationship slowly, balancing their trust in you with your ambitions for their support.

15.5.6 Miscellaneous possibilities that really ought to be mentioned somewhere

- Multiple/matched funding: Two or more funders, and perhaps your institution participating financially.

- Seed funding: Initial investment to start a small company to exploit a new idea (e.g. crowdfunding).

- Venture capital: Private equity (share ownership) funding to take forward a seed-funded success.

- Spin-outs and start-ups: Small companies established to exploit a new idea with venture capital.

- Business incubators: Companies that help start-up companies with training or space.

15.6 Health service research in the UK

Arrangements for funding of research directly relevant to clinical practice in the NHS is different in each of the four UK countries. The largest and most well developed is the National Institute for Health Research (NIHR) in England, which is described briefly in this section and in further detail in Chapter 16. There is no exact equivalent to the NIHR in the other three UK countries, however, research funding is provided in Scotland by NHS Research Scotland and in Wales by Health and Care Research Wales. In Northern Ireland, the Public Health Agency – Health & Social Care Research and Development provides information about funding opportunities that are available to researchers working in Health & Social Care (HSC) Northern Ireland but does not itself provide research funding.

The National Institute for Health Research (NIHR), established in 2006, is funded through the Department of Health in England to improve the nation's health through research, and claims to be the most integrated clinical research system in the world. The NIHR manages its health research activities through four main work strands.

- **Infrastructure:** providing the facilities and people for a thriving research environment.

- **Faculty:** supporting the individuals carrying out and participating in research.

129

- **Research:** commissioning and funding research.

- **Systems:** creating unified, streamlined and simple systems for managing research and its outputs.

It's no coincidence that many topics of research that emerge within the health service itself have a natural affinity with one of the elements of the NIHR's research strands, which are shown in detail in Table 15.2.

Many successful researchers began their careers with the support of the NIHR and your attention is drawn to the specific activities listed in Table 15.3.

Table 15.2 NIHR research strands

Strand	Scope	Units	Detail
Research	Commissioning and funding research	Schools	• School for Primary Care Research (SPCR) • School for Public Health Research (SPHR) • School for Social Care Research (SSCR)
		Programmes	**Research Programmes** • Efficacy & Mechanism Evaluation (EME) Programme in partnership with the MRC • Health Service & Delivery (HS&DR) Programme • Health Technology Assessment (HTA) Programme • Methodology Research Programme, in partnership with the MRC • Invention for Innovation (i4i) Programme • Programme Grants for Applied Research (PGfAR) • Programme Development Grants (PDG) • Public Health Research (PHR) Programme • Research for Patient Benefit (RfPB) Programme • Systematic Reviews Programme (SRP) • Cochrane Review Groups (CRG) • Technology Assessment Reviews (TAR) **Other Programme Work** • Blood and Transplant Research Units (BTRU) • Centre for Surgical Reconstruction and Microbiology (CSRM) • Clinical Trials Units (CTU) • Health Protection Research Units (HPRU) • Horizon Scanning Research & Intelligence Centre (HSRIC) • INVOLVE • James Lind Alliance Priority Setting Partnerships (PSP) • PROSPERO • Research Design Service (RDS) • UK Cochrane Centre (UKCC)

NIHR: From Structure to Funding (assembled from data at www.nihr.ac.uk/)

Table 15.3 Activities of specific strands within the NIHR

Research Programme	Significance
Health Technology Assessment (HTA) Programme	The HTA funds research about the effectiveness, costs and broader impact of healthcare treatments and tests. A wide range of methods are applicable.
Invention for Innovation (i4i) Programme	i4i is a translational scheme to advance healthcare technologies for patient benefit in areas of clinical need. See also the Health Innovation Challenge, a partnership with Wellcome.
Programme Grants for Applied Research (PGfAR)	PGfAR produce independent research findings with practical benefit. Programme *Development* Grants (PDG) are a complementary preparatory scheme.
Research for Patient Benefit (RfPB) Programme	RfPB is a national, response-mode programme to generate high quality research for the benefit of NHS users. Its purpose is to achieve, through evidence, the potential for improving the way that healthcare is delivered for patients, the public and the NHS.

15.7 Practicalities – logical steps to a funded research career

15.7.1 The first small step

Get on the funding ladder as soon as you can. Funders like to support those who are already successful – they take reassurance that you are a safer bet. To be able to quote funding success, you need an easy first target, so seek support for your travel to – and participation at – conferences, for small items of equipment, and for dissemination and knowledge transfer. It's never too soon to start, and it's perfectly possible to achieve success even during your training or PhD. It enables you to quote previous funding on your next application, and on your CV, and it is an early indicator of persuasive writing ability.

15.7.2 Help from institutional expertise

Your manager or PI are obvious first points of contact. But talk to grant winners, and to your trust or university research office. Some truly visionary institutions have appointed business development managers, who have significant expertise and experience and are a joy to deal with. Importantly, you should not try to go it alone, you will almost certainly perish.

15.7.3   A first research grant

If you're an academic, the Research Councils have created a special scheme for you – the First Grant (or 'New Investigator') Scheme. It requires serious effort to construct a winning proposal, and the statistics aren't good: in 2014/15, the MRC received 144 applications and funded 24, just 17%. Clinical fellowships do better, with around 30% being successful, but non-clinical do worse. As a consequence, many academics, and many NHS personnel, instead consider the smaller charities and the NIHR.

15.7.4   A portfolio: equines for circuits

Once you've broken the funding surface ice, unless you're a serial EU funding wizard, you need to consider avoiding being associated with just a single funding approach. The researcher funded four times by the same small, niche charity is clearly frightened to take the next step. The pathway is clear – bigger, bolder, but probably heading for a collaboration. You may already be part of a clinical/academic mix, but you need to consider a spectrum, not least because different parts of the journey require different funder specialisation. A landscape well known to IPEM members is the development of computational biomarkers to enhance clinical decisions. The early work may be funded by a specialist charity, later a consortium will take it to RCUK or the EU, a clinical trial will need the Wellcome Trust or maybe the NIHR, and introduction to the market will be a commercial collaboration. Then there is the regulators... but it's all possible. And there might be some intellectual property to make it all worthwhile.

15.7.5   The special cases

We're all special cases – that's why personalised medicine will conquer the world. But in research, there are some activities that always need the same treatment. An example is i4i (Invention for Innovation – a translational funding scheme) where, if you work in a territory that consistently converts clinicians' bright ideas into electromechanical reality, your pathway is clear. But it's probably time to lead that clinician into a new experimental territory and fly higher.

15.7.6   Industry

Almost nothing goes on to make a difference in healthcare without the involvement of a commercial organisation; even the rearrangement of a care pathway turns out to need a new computer system to run it. So working with industry needs to become a way of life. As soon as you can, identify the likely beneficiaries of your expertise, and start a conversation; if necessary get help from an institutional link. Companies are your friends, they focus your attention, they bring you sustainability and impact, and ultimately they deliver improved care to patients.

15.7.7 Gateways, innovation hubs, technology sector growers...

Healthcare has always been a growth industry, and there is an associated growth of interface organisations trying to reduce the friction. Get to know your local agencies in these categories, understand their relative strengths and values, and talk to them about partnerships, intellectual property, exploitation (of the ideas...) and the onward journey. They only bite when you get boring.

15.8   Recapitulation

To conclude, here is a tangential look at the topics covered in the main text, suggesting the qualities that, when cultivated, will make your projects fundable and your research career a triumph.

15.8.1   Temperament

Successful medical research requires a combination of skill, knowledge, training, curiosity and determination. It probably also needs an ability to work with others so that cooperation can magically produce more than the sum of the individual contributions. If you weren't born with a temperament that supports this combination of attributes, reflect carefully on whether it will bring you the satisfaction required to maintain your interest. Reluctant research is probably suboptimal research.

15.8.2   Imagination, inspiration, inventiveness... investment

The additional quality that leads to success is perhaps the ability to look at a problem and see within it the opportunities to structure, design and unify a set of investigations that will yield an achievable goal. This chapter has outlined the inevitability that, unless these investigations can additionally be polished until they are attractive to funders, the ambition will remain frustrated, so developing a special type of inventiveness – one that automatically carries a sense of *fundability* – is a vital attribute.

15.8.3   'Terroir' and trajectory

The French concept that an output is inevitably affected by every one of the inputs that came into play during the construction process is usually taken to refer to agriculture, particularly wine production. But a similar concept pervades the research process, and an awareness of your own overall environment (thinking, working, clinical, research, commercial...) is important to the staging of the individual steps in the process, and to the selection of the appropriate funders. There are multiple factors contributing to your research life and, taken together, they represent the 'terroir' that is your own unique domain, that will flavour your own results. As you map out this working territory, so

133

the cluster of funders that fit the different aspects of this profile will be identifiable. Get to know these funders as well as possible, talk to other recipients of their funding, understand the funders' relevance to the steps in your developmental journey. And it's a cooperative process – funders need to demonstrate success just as much as you do.

15.8.4   Skill mix

The ability to construct a fundable scenario is the magical element, but there are other skills that you need to develop.

Thinking: Critical thinking is probably a native skill, and you're interested in research because you've already been unable to prevent it kicking-in. Now you need to hone this skill to perfection, to have an answer for every critic, to demonstrate that your way is the best way, and to make clear that you've thought of all of the alternatives, or at least all those the reviewers can come up with. Question everything (Why?).

Writing: It's not enough to have a good idea, you must have the ability to communicate it to others. Read successful proposals, and talk to their authors. Consider the resources identified earlier. Practice writing, develop re-usable text appropriate to your field, lay out your arguments and rearrange them until the sequence is unbeatable. Pay attention to reviewers' comments, and read on and between the lines.

Networking: Serious researchers tend to have many contacts, not necessarily because they are naturally gregarious, but because their success has expanded their collaborations, and the need for collaboration has driven their success. Get to know the others who are prominent in your field, and in associated fields, and force yourself to make contact with them.

15.8.5   The holy grail?

The grail promises infinite abundance, so may be a tad ambitious, but winning research funding is a significant step. Many researchers feel the need to justify themselves by seeking ambitious support from prestigious and heavily oversubscribed sources. Such an approach requires dogged determination, patience, and a thick skin; more importantly, in the time it takes to win Research Council support, a typical researcher could probably have won three grants from small charities, and established a track record in their domain. Early, low-level success will not count against your future prospects; on the contrary, the sooner you achieve results, the sooner you can take your second step. Don't hesitate – jump in, the water's lovely.

Appendix 15.1 Resources

How To Be A Researcher

http://www.jobs.ac.uk/careers-advice/working-in-higher-education/1203/5-skills-you-need-to-become-a-researcher

http://www.itworld.com/article/2749753/careers/career-planning--so-you-want-to-be-a-researcher.html

https://www.vitae.ac.uk/researcher-careers/pursuing-an-academic-career/research-funding

http://careers.bmj.com/careers/advice/Research_options_for_doctors_in_training

Funding Notifications

https://www.vitae.ac.uk/researcher-careers/pursuing-an-academic-career/research-funding/where-to-find-sources-of-academic-research-funding

http://www.researchresearch.com/

https://www.researchprofessional.com/sso/login?service=https://www.researchprofessional.com/0/

http://www.idoxgrantfinder.co.uk/

http://www.rdlearning.org.uk/

Proposal Writing

https://www.epsrc.ac.uk/funding/howtoapply/preparing/

http://www.insight.mrc.ac.uk/2015/10/05/12-top-tips-for-writing-a-grant-application/

https://www.freelancer.co.uk/

Consortium Formation

https://blog.piirus.ac.uk/2015/08/11/5-ways-to-find-collaborations-that-further-your-research/

http://www.jobs.ac.uk/careers-advice/working-in-higher-education/1732/academic-research-a-guide-to-building-international-collaborations-for-academics

http://www.nihr.ac.uk/about-us/how-we-are-managed/our-structure/infrastructure/collaborations-for-leadership-in-applied-health-research-and-care.htm

http://gw4.ac.uk/guidetoresearchcollaboration/getting-started/finding-collaborators/

General Advice

http://www.womeninresearch.org/

http://www.jobs.ac.uk/careers-advice/working-in-industry/1957/research-roles-within-the-nhs

National Bodies

http://www.rcuk.ac.uk/

http://www.crac.org.uk/

Charitable (starting points)

http://www.amrc.org.uk/

http://www.health.org.uk/

http://www.charitychoice.co.uk/charities/health

Commercial

Medical Imaging: Fujifilm Holdings, GE Healthcare, Siemens Healthcare, Philips Healthcare, Shimadzu Corporation, Toshiba Medical Systems Corporation, Carestream Health, Hitachi Medical Corporation, Hologic, Esaote

Medical Equipment

Non-imaging: Johnson & Johnson, Baxter International Inc., Fresenius Medical Care AG & Co., KGAA, Koninklijke Philips NV, Cardinal Health Inc., Novartis AG, Covidien plc, Stryker Corp., Becton, Dickinson and Co., Boston Scientific Corp., Essilor International SA, Allergan Inc., St. Jude Medical Inc., 3M Co., Abbott Laboratories, Zimmer Holdings Inc., Terumo Corp., Smith & Nephew plc, CareFusion Corp., Getinge AB, Olympus Corp, Bayer AG, CR Bard Inc., Varian Medical Systems Inc., DENTSPLY International Inc., Ship Healthcare Holdings Inc., Paul Hartmann AG, Nipro Corp., Coloplast A/S, Sonova Holdings, Danaher Corp., Edwards Lifesciences, Intuitive Surgical Inc., MIRACA Holdings Inc., Dragerwerk AG & Co. KGa

Gateways, Innovation Hubs, Sector-growers

http://www.nesta.org.uk/

http://shg.sheffield.ac.uk/

http://www.medipex.co.uk/

http://medilink.co.uk/

Intellectual Property

https://www.gov.uk/topic/intellectual-property

http://www.wipo.int/about-ip/en/

http://www.amrc.org.uk/publications/benefiting-innovation-intellectual-property-advice-medical-research-charities

Appendix 15.2 Major UK Research Funders

The major UK research funders are listed below.

Research Councils

Council	Web
Arts and Humanities Research Council	http://www.ahrc.ac.uk/
Biotechnology and Biological Sciences Research Council	http://www.bbsrc.ac.uk/
Engineering and Physical Sciences Research Council	http://www.epsrc.ac.uk/
Economic and Social Research Council	http://www.esrc.ac.uk/
Medical Research Council	http://www.mrc.ac.uk/
Natural Environment Research Council	http://www.nerc.ac.uk/
Science and Technology Facilities Council	http://www.stfc.ac.uk/

Academies

Academy	Web
British Academy	http://www.britac.ac.uk/
Royal Academy of Engineering	http://www.raeng.org.uk/
Royal Society	https://royalsociety.org

Major Medical Research Charities

Charity	Web
Arthritis Research Campaign	http://www.arthritisresearchuk.org/
British Heart Foundation	https://www.bhf.org.uk
Cancer Research UK	http://www.cancerresearchuk.org/
Leverhulme Trust	https://www.leverhulme.ac.uk/
NC3Rs	https://www.nc3rs.org.uk/
Nuffield Foundation	http://www.nuffieldfoundation.org/
Wellcome Trust	https://wellcome.ac.uk/

Further details of medical research charities can be found from the Association of Medical Research Charities (AMRC); http://www.amrc.org.uk

Funding bodies relevant to the National Health Service (UK)

Area	Funding body	Web
England	National Institute for Health Research	http://www.nihr.ac.uk/
Scotland	NHS Research Scotland	http://www.nhsresearchscotland.org.uk/
Wales	Health and Care Research Wales	https://www.healthandcareresearch.gov.wales
Northern Ireland	Public Health Agency – Health & Social Care Research and Development *	http://www.hra.nhs.uk

*Note that the PHA-HSCR is not a funding agency, but provides information about funding opportunities that are available to researchers working in Health & Social Care (HSC) Northern Ireland.

16 Research and innovation in the NHS

Wendy B. Tindale

16.1 Introduction

In the last 10 years, the climate in the National Health Service has changed almost beyond recognition. There has been a fundamental restructure, huge steps forward in terms of technological innovation, a reframing of what it means to be a caring professional following the Francis enquiry (www.midstaffspublicinquiry.com), and a focus on patient-centricity and care closer to home. All this has been against a growing backdrop of financial uncertainty and affordability. The research culture in the NHS has also changed, despite, or possibly because of, the broader context. Research and innovation are now seen as being key to delivering transformation and sustainable change; indeed the Health and Social Care Act of 2012 (www.legislation. gov.uk) identified the NHS Commissioning Board (now NHS England) as having a duty to promote both research and innovation in matters relevant to the provision of health services.

That research and innovation are welcomed in today's NHS is not in doubt, but the context is substantially different to what some still see as the 'good old days'; when anyone could 'have a go at a bit of research', where the costs of research could somehow be lost in the system and not accounted for, and where research approval simply meant your boss saying it was OK. Today's NHS research environment is serious business and it extends far beyond the major teaching hospitals. Research is perceived as a core activity; it is transparent and fully accountable, with a national infrastructure and a performance-based approach to funding. It is also collaborative, in the same sense that university-based research is now much more collaborative, with the most successful research now being delivered by large, sometimes geographically disparate teams – the age of the lone researcher has long since passed. But it is also highly competitive and the NHS research activity league tables (www.nihr.ac.uk/ research-and-impact/nhs-research-performance/league-tables/) are a coveted measure of esteem.

Never has there been a better time for budding researchers in healthcare: opportunities abound for those who are able and committed. This chapter describes the context that researchers need to be aware of and the support and funding streams that are available, and provides advice about how to get started and some of the key considerations that need to be addressed.

16.2 The National Institute for Health Research

There is little doubt that the National Institute for Health Research (NIHR) has been a catalyst for major change. Created in 2006 and funded by the Department of Health, its vision is

> *to improve the health and wealth of the nation through research*

and its mission is

> *to provide a health research system in which the NHS supports outstanding individuals working in world-class facilities, conducting leading-edge research focused on the needs of patients and the public*

It has a budget in excess of £1 bn. Before the creation of NIHR, funding for research activities in the NHS was *ad hoc*, monies were not necessarily allocated to areas of research need and there were capacity and operational challenges in delivering research activity. Today, there is a clear national research framework, with transparency, quality and impact assessment as core principles (www.nihr.ac.uk).

The NIHR is often termed 'the research arm of the NHS'. To the uninitiated, its structure and myriad of programmes can seem complex, but time spent understanding what is on offer will pay dividends and is certainly recommended for the novice health researcher. Essentially, the NIHR has four main strands of activity, covering the funding of:

    i.   research projects and programmes themselves;

    ii.   infrastructure (people and facilities) to support research;

    iii.   individuals undertaking research (including research training); and

    iv.   centralised research management systems.

The first three of these are likely to be most relevant to those embarking on a health research career. There are a number of research project funding schemes, which are nationally competitive, many of which have both themed commissioned calls and open calls. Of particular interest, though, are the infrastructure organisations and programmes, as these can often be a useful entry point for the less experienced researcher. Examples include Biomedical Research Centres (BRCs), Experimental Cancer Medicine Centres (ECMCs), Medtech and *in vitro* diagnostic Co-operatives (which from 2018 will replace the Diagnostic Evidence Co-operatives and Healthcare Technology Co-operatives (DECs and HTCs)), Collaborations for Leadership in Applied Health Research and Care (CLAHRCs) and Patient Safety Translational Research Centres (PSTRCs). All of these have a different focus but share a common aim to provide the support and facilities the NHS needs to underpin its research. It is through these types of organisations that those wishing to pursue a career in research can often find opportunities, perhaps to join a research team, collaborate with others or receive training.

The NIHR also funds a Research Design Service (RDS) which exists to provide support for research study design and methodological approaches. It supports NHS researchers and those working in partnership with the NHS, for example, university staff or researchers, and is intended to help those developing an application for funding. It can be invaluable for those who are less experienced in crafting an application or who need guidance in terms of what a funding panel would be expecting to see in an application. The RDS operates through 10 regional centres which cover the whole of England (www.rds.nihr.ac.uk).

The NIHR funding schemes support health research in England and, usually, also Wales (but note that not all schemes are open to applicants from Wales and it is advisable to study the eligibility criteria carefully). Some schemes require the award to be administered by an NHS organisation, although the applicant does not need to be an employee of that organisation providing they have an appropriate relationship (for example, an honorary contract). This is not the case for all schemes – in some, the lead organisation may be a university, an NHS organisation or a company. The best advice is to study the 'Frequently Asked Questions' which are relevant to the grant scheme of interest.

For those in Scotland and Northern Ireland, the nearest equivalents to the NIHR are NHS Research Scotland (www.nhsresearchscotland.org.uk), which is a partnership between the Scottish Health Boards and the Chief Scientist Office of the Scottish Government, and the Health and Social Care Research and Development Division (www.research.hscni.net) which is part of the Public Health Agency in Northern Ireland.

16.3   Scientists in the NHS

Healthcare scientists in the NHS have a key role in shaping future services and work at the forefront of science and technology. Their training provides a sound basis for leadership in applied research and innovation. In addition, being embedded in the NHS means that they have first-hand experience of the needs of the NHS and its patients. They have the opportunity to lead healthcare developments and to generate evidence about impact and they have an important role in the adoption of new research outcomes, technologies and systems. With the current emphasis on data management, analytics and precision medicine, never before has there been such a need for physics and engineering-based disciplines in healthcare. But partnership between scientists and engineers working in the NHS and those in academia is crucial if we are to make more than incremental changes to the way in which health and care are delivered; disruptive technologies which can underpin transformational change are more likely to arise from innovative partnerships which can create 'out of the box' thinking. Increasingly, research partnerships with industry are more commonplace, as we move from transactional to collaborative relationships with relevant industries. The message is clear – healthcare scientists must look externally as well as internally for partners, and academics must collaborate with the NHS in order to deliver genuinely impactful research.

The terms *research* and *innovation* are often used in a rather imprecise way in relation to healthcare. In fact they are part of the same spectrum and usually both are necessary to create impactful change. The Cambridge English Dictionary (www.dictionary.cambridge.org) defines research as

> *A detailed study of a subject, especially in order to discover (new) information or reach a (new) understanding*

It defines innovation as

> *(the use of) a new idea or method.*

The NHS (2011) has defined innovation as

> *An idea, service or product, new to the NHS or applied in a way which is new to the NHS, which significantly improves the quality of health and care wherever it is applied*

All scientists in the NHS are expected to innovate, but not all will be involved in research.

Where NHS scientists are undertaking research, it is important to recognise that this may not involve clinical trials; indeed, it may not directly involve patients. Much of the research performed by physicists and engineers may be bench-based or computer-based, for example, involving technique or device development or data analysis. However, this activity should still be subject to the relevant research approvals processes and be formally identified as such.

Irrespective of whether an NHS scientist is involved in research or innovation, it should be incumbent upon them to disseminate their outputs, for the benefits to patients and to inform the broader healthcare community. The dissemination route will vary depending upon the context and will range from publication in peer-reviewed journals (usually an expectation for formal research) to information-sharing workshops, facilitated events and the use of a variety of media to promote sharing of best practice and novel ideas with proven benefit.

It is worth noting that, in addition to the NIHR funds for research, NHS England (the commissioning body for healthcare services in England) separately funds innovation activities, usually through competitions, and this can be a valuable source of money to support local improvements in services or the transition to new models of care.

16.4   Get advice and get networked...

If you have a new idea which could be applied in healthcare, it is important to be clear on what is needed to get started. Is it research? Innovation? Is it simply an improvement to your 'business as usual'? Many organisations will have departments which can provide advice, for example, a research office or a research and innovation service, and some will have online toolkits which can help guide you through that early decision-making. Conducting research will almost certainly require an organisational approvals process and, depending upon the nature of the research, may require ethical approval. Local research offices are usually the main source of advice. If you are applying for research funding, a conversation with the local RDS is a must; indeed, many funding streams will want assurance that an RDS has been involved at the planning stage. The RDS can also provide an interface to other relevant organisations – a Clinical Trials Unit, for example, if you are planning to undertake a clinical trial, or the Clinical Research Network, should you be undertaking a multicentre trial or perhaps a commercially sponsored study. You should take advantage of this support – it can often mean the difference between success and failure of a funding application, even if you are an experienced researcher.

Being networked as a researcher is also an important part of the journey to success. Network with those in your field locally, nationally and where possible internationally, and ensure that you are aware of what is state-of-the-art in your area. Whilst research journals are a key source of knowledge, 'softer intelligence' is also of value so ensure you are aware of, and join, relevant professional and other groups. For example, there may be mailbases which provide useful information, or LinkedIn groups which will ensure you remain connected.

Take time to understand the research and innovation landscape in your area of interest. There may well be organisations which can provide useful advice and which are funded to do so. If your research involves the development of a new technology, a diagnostic technique or an intervention, there could be intellectual property considerations. It is important to know what is already on the market, what the future market might look like for your new development, and whether you have 'freedom to act' – is there a patent in existence which could block your development, for example? There are technology transfer offices and innovation hubs in Universities and supporting the NHS which can advise. The Horizon Scanning Research and Intelligence Centre (www.hsric.nihr.ac.uk) provides information on emerging technologies in healthcare, which is another useful source of knowledge. Do you know the key research questions in your area? The James Lind Alliance (www.jla.nihr.ac.uk) has been established to identify the top 10 priorities for research in a multitude of different clinical areas and the National Institute for

Health and Care Excellence (NICE; www.nice.org.uk), as part of its published guidance, will often highlight areas where more research is needed. In addition, NICE also publishes MedTech Innovation Briefings. These are intended to support health and care commissioners who are considering using new medical technologies, but they are a valuable resource for those in the medical technology development field, as they provide information on up and coming technologies which might otherwise be difficult to find.

Don't forget that the HTCs and DECs may be relevant to your research endeavours. The eight HTCs are national centres of excellence that work collaboratively to gain an in-depth understanding of unmet needs and develop concepts for new medical devices, healthcare technologies and technology-dependent interventions that improve treatment and quality of life for patients. The four DECs help to generate evidence on the clinical and cost-effectiveness of *in vitro* diagnostics, helping to improve the way diseases are diagnosed. Both types of organisation provide the opportunity for new, interested researchers in a relevant field to get involved and they may provide a nurturing pathway to inclusion in a first grant application. From 2018, the NIHR will replace the current HTCs and DECs with Medtech and *in vitro* diagnostic Co-operatives (MICs). These will incorporate and retain the remits of both schemes and will form a single NIHR infrastructure scheme to support medical technologies and commercially supplied *in vitro* diagnostics.

16.5   Need money to get started?

It is difficult to undertake serious research without funding. But getting funding often requires that you are an established researcher with a track record of successful delivery. In addition, funding bodies usually expect to see some proof of concept work and rarely fund a new idea without some feasibility data.

The easiest way into funded research is to work with an established research group which is already funded for a programme of related research activity. Networking is important here, so use your contacts to find relevant groups and ensure that you can play a useful role and add value to the team. Research groups are often looking for 'pairs of hands' to undertake short projects which will provide pilot data to support a new avenue of exploration, or someone with literature searching or data analysis capability, and this can be a useful way in, particularly if you have a background in the area.

Many organisations have small pots of 'pump-priming' funds, which can be easier to access than nationally competitive funding schemes. These may be local charitable funds, proof of concept funds, 'bright ideas' funds, etc., and may cover both research and innovation. It is worth finding out what is available in your local area and seeking advice in terms of what is most relevant to your needs. Each fund will have terms of reference covering the uses to which the monies can be put and applicant eligibility criteria, so find out what these are and how you fit within them. Applying to these types of funding schemes will not only enable you to get started, but it will also demonstrate that you are serious about research and understand the importance of identifying funding. It will be a valuable addition to your CV if you are successful.

There may also be local opportunities for buying out some of your time to enable you to have protected time for research. These usually apply to NHS staff, where they have a service commitment which needs fulfilling. Schemes allow for temporary backfill arrangements to enable the applicant to have protected research time for a fixed duration, which may be full-time or part-time. Sometimes, such schemes are focused on enabling the individual to have time to complete a grant application, supported by the RDS, with the intention of providing a route to sustainable research if the grant application is successful.

As noted above, NHS England funds innovation activities. These tend to be announced in support of new policies or initiatives and are less predictable than the regular rounds of NIHR funding, but may be a good fit for service innovation. Keep a watching brief on NHS England innovation webpages and engage with organisations such as the Academic Health Science Networks (AHSNs; www. ahsnnetwork.com) in order to become alerted to upcoming innovation funding calls.

It is important to be aware of all of the different funding sources which may be relevant to you: Research Councils, Wellcome Trust, Health Foundation, etc., in addition to those mentioned above. It is noteworthy that the Institute of Physics and Engineering in Medicine (IPEM; www.ipem.ac.uk) offers research and innovation awards which fund the purchase of equipment or services to facilitate short-term projects. One of the easiest ways to maintain awareness is by registering with websites or departments which provide regular circulars highlighting new funding calls. Academic finance offices and research offices are a useful source of advice.

16.6   Develop a strategy

What do you want to achieve? Are you looking to pursue a burning idea or do you have a longer term plan to become a clinical academic? It is important to consider this carefully because the pathway may well be different in terms of your chosen option.

If you are looking to pursue an idea, then the preferred route may well be through linking with an established research team and building your credibility so you can apply for a research grant to develop your idea. If your desire is for a clinical academic career, then you may wish to consider the Health Education England (HEE)/NIHR Integrated Clinical Academic (ICA) Programme for non-medical healthcare professionals.

Clinical academics are usually identified as healthcare professionals whose time is divided between their clinical practice and academic work involving research and the delivery of education. The term was originally used for medical doctors who were university employees, spending roughly half their time as practising doctors and the other half on research and education. However, it is now used more generically to refer to any healthcare professional who has a formal dual role split between practice and research. Clinical academics can be university employees with an honorary NHS contract, or alternatively, they may be NHS employees with an honorary university contract.

It is important to note that there are university employees who are engaged in healthcare research activities, who may well have an honorary contract with an NHS organisation to facilitate collaborations, but who are not clinical academics in the formal sense. The key differentiator for a clinical academic is whether the individual is involved in clinical practice as part of their role, for example, this might involve working in an NHS clinical engineering role or an NHS imaging or radiation scientist role for part of the time.

The ICA Programme for non-medical healthcare professionals provides personal research training awards for eligible healthcare professionals who wish to pursue a career which formally combines research and clinical practice. They include Clinical Doctoral Research Fellowships, Clinical Lectureships and Senior Clinical Lectureships. The awards are made annually through a competitive process and require the support of both NHS and academic host organisations. Consideration must be given to both academic and clinical mentorship/supervision and it is worth considering the use of a range of mentors, ideally from more than one organisation, in order to avoid appearing too 'insular'. These programmes are intended to support both research and research training, as well as more advanced development of clinical expertise. The development of research leadership is important and applicants are encouraged to participate in leadership support and development programmes, recognising the important contribution this can make to research performance and to the growth of future research leaders. Anyone applying for the ICA Programme would be well-advised to consult their local RDS; preparing a good application can take anything up to a year when you have other commitments as a full-time professional, so starting the process early is essential.

At the present time, there are a very limited number of clinical academics in medical physics and clinical engineering, but with the advent of the ICA Programme and the opportunities for progression within it, there is an anticipation of growth.

At the pinnacle of research awards are the NIHR Research Professorships. These will fund outstanding academics who are not yet research leaders, but who are at consultant grade or equivalent. The funding is for a 5-year period and is intended to consolidate research leadership, particularly in translational research in defined fields, which include experimental medicine and methodological research.

146

16.7   Adding value

Make sure that what you want to do in the research and/or innovation sphere will add value to your organisation. Do you know your organisation's goals and strategy and is there a good fit with your ideas? If there is not, it will be difficult to gain the organisational support that you will need, so consider carefully how to frame your ideas and the context in which you are operating. It is a good idea to discuss ideas at an early stage with someone in a senior position who can advise on where you might need support or buy-in and help to consolidate the way in which you might shape your thoughts to maximise the contextual fit and value to the sponsoring organisation (your NHS employer).

Research and innovation which has a clear fit with needs, be those patient or service needs, and where benefits will be measurable and linked with SMART objectives (www.cdc.gov/HealthyYouth/evaluation/pdf/brief3b.pdf), is more likely to receive backing. Equally, NHS Trusts recognise the importance of the research activity league tables. It is a good idea to familiarise yourself with the metrics (www.nihr.ac.uk/research-and-impact/nhs-research-performance/league-tables/) which are used for the league tables and to ensure that, wherever possible, your proposed activity will assist the organisation in delivering against these metrics.

If you are an NHS researcher undertaking collaborative research, it is possible that you may need an honorary contract with a university in order to work across organisational boundaries with all of the necessary approvals and governance arrangements in place. The equivalent applies if you are a University researcher. Human Resources departments or research offices can help to point you in the right direction. Whenever you are working in collaboration, give some thought to how you can add value more broadly to all of the participating organisations. Sometimes, this may be as simple as ensuring you give appropriate credit in your dissemination activities – enabling all partners to leverage value and have some ownership of the outcomes. Whatever the approach, keep this in mind – it matters and you will get far more support and opportunity by giving this aspect some attention.

16.8   Don't forget the patients and the public!

Patient and public involvement (PPI) is an absolutely key part of successful research. It is now an expectation of most, if not all, funding bodies that there is public involvement in research and indeed, without this, it is extremely difficult to be successful in nationally competitive schemes. INVOLVE (www.invo.org.uk) is an excellent resource which helps guide individuals looking to involve members of the public in their research.

But this is not just about paying lip service to the requirements of funding bodies. Involving patients and the public makes the research richer, more focused and more impactful. How do you really know what patients need unless you ask them? How do you know what they would find most helpful? What would end up being a barrier? What would be unacceptable? What would engage or stop people from entering your clinical trial? These types of questions are readily addressed if you embrace PPI. Don't make the mistake of thinking that we are all patients at one time or another and therefore you and/or your colleagues can act as a proxy. PPI is not necessarily easy and can be a little scary to the new researcher, but there is plenty of support to draw upon. Speak to the NIHR infrastructure organisations in your local area for advice, they may well have patient panels already in existence who would be pleased to offer their expertise. Patient governors in NHS Trusts are another useful avenue for exploration, as are charities.

Patient and the public can be involved at all stages of research: setting the research priorities, advising on recruitment, helping with research applications (particularly the plain English/lay summaries which are usually required), participating in steering committees, acting as advocates, and assisting with dissemination activities. The patient voice can be very powerful when it comes to the translation and adoption of research outcomes and can often help to overcome barriers which might otherwise exist.

As a researcher in today's environment, it is important to make PPI a priority. Even if your research is laboratory- or computer-based or bench-top, do not think that PPI does not apply – it does! PPI is not free, and you should consider how to provide payment and recognition for public involvement, including travel expenses. Often, members of the public are happy to contribute for free, but this should not be an expectation. Guidelines on payment can be found on the INVOLVE webpages. If you are conducting pilot or feasibility work, it may be that you have no dedicated funds that you could use for this. However, there are a number of small pots of funding which can be accessed for this purpose, including small Trust funds, local charities and RDS monies. Use this as an opportunity to seek out your first grant – a few hundred pounds will usually be sufficient.

16.9   Conclusions

Research in the NHS has entered a new era. The NIHR, as the research arm of the NHS, has changed both the culture and the opportunity for research. NIHR funding is a coveted badge, not only for NHS researchers but also for academics engaged in health-related research.

There is now plenty of formal research training on offer – albeit through competitive schemes – and plenty of support to enable novice researchers to get started.

Collaborations, across many different sectors and organisations, be they with academia, industry, charities or the public, are now more important than ever, and are key to the delivery of successful and impactful research.

The research and innovation agenda within the NHS is seen as critical for the service transformation which is required for future sustainability and there is a need for scientists and engineers to understand the needs and create the technologies and interventions which can be part of that change.

There is a huge opportunity – embrace it!

References

NHS (2011) Innovation, health and wealth; Accelerating adoption and diffusion in the NHS. Available at http://webarchive.nationalarchives.gov.uk/20130107105354/ http://www.dh.gov.uk/en/Publicationsandstatistics/Publications/PublicationsPolicy AndGuidance/DH_131299

17 Funding and managing research and development in the industrial sector

Patrick Finlay

Once a new idea has been demonstrated in lab conditions, maybe at the end of a PhD or a funded academic research project, the prospect of developing a commercial product becomes an interesting coffee-break discussion. But what is involved in moving a conceptual medical device 'from lab bench to bedside'? What do we let ourselves in for if we set off down the poorly signposted path to global marketing success?

17.1 Commercial research is a different ball-game

The change of culture between academic and commercial environments should be approached with eyes open. It can be unexpectedly different. The purpose of academic research is to expand knowledge and stretch the limits of what is possible. But commercial research is tightly defined: the intention is to specify and develop the optimum product for meeting a particular user need.

Commercial research into a new idea starts by asking what is the user's problem that this technology can solve. In fact, the User Requirements Specification (URS) is the key document that guides a commercial R&D programme. If the user wants something that can move at 2 m.s^{-1}, the researcher must learn to stop when this goal is achieved. It may be professionally satisfying to spend another few months honing the speed to a record-breaking 2.2 m.s^{-1}, but in the focused world of commerce, that is wasted time and effort. In the same way, the researcher's natural instinct is to make each device he builds slightly better than its predecessor, but the marketing director would like them all to be the same, for very sound reasons. For example, it is hugely difficult to provide after-sales support if every device is different, and if customers get to hear that the next machine will be an improvement, they are likely to hold off purchasing. Commercial discipline that focuses on the URS goals is part of the ethos, and it is one of the hardest changes for an academic researcher to adjust to.

The other cultural shift is from the phenomenal to the consistent. It's often a triumph in PhD research to demonstrate that your idea works once, even though success may involve many failed attempts and require near-ideal conditions. But in commercial R&D, the measure of success is to achieve *dependability*, a catch-all term that embraces safety, reliability, maintainability, availability and legibility ('the psychological ability of a user to understand what the device is doing'). Dependability to the level set in a typical URS is a demanding goal, and usually requires much

150

patient analysis of why things aren't working perfectly every time. This is called 'development', and it is different from 'research'. It means isolating and controlling the variables until a consistent performance can be guaranteed throughout the specified range of environmental conditions.

Many medtech companies employ a 'breaker' – someone whose job it is to try to make a new product fail by subjecting it to tortuous conditions of use. Beating the breaker is where good commercial R&D engineers derive their happiness.

The reality that surprises many academics is that their university research was the easy bit: the development of a new product becomes progressively more complex and expensive as it gets nearer to commercial launch. As a rule of thumb, if academic research to demonstrate lab bench feasibility costs 1 unit, commercial R&D to produce a regulatory approved prototype will cost 10 units, and the final productionising and industrial engineering of a commercial product will cost a further 5–8 units. And for a large company, all of this is likely to be dwarfed by the cost of a global marketing launch campaign.

17.2   Starting off

It may be possible to take your new idea directly to an existing manufacturer who will license it from you, especially if your contribution is something like a new algorithm that can be readily demonstrated to offer an improvement. But more usually if you want to develop a new medtech idea arising from academic research, the first step is to set up a company. This provides a vehicle for funding the project, owning the patents, employing the entrepreneur(s) and trading goods and services. The company's website and branding give the project visibility and credibility.

The UK Medtech sector consists of around 3300 companies, nearly all of which are small with less than 50 employees and less than €10M turnover. A significant number are young high-tech spin-outs from universities, started by one or more entrepreneurial graduates sometimes underwritten by their university's enterprise fund.

The formalities of setting up a UK limited company are very easy. It can be done online via the Companies House website (www.gov.uk/government/organisations/companies-house). More complex are the arrangements for employing staff: you will benefit from legal advice from a commercial solicitor for wording contracts, and for complying with a considerable volume of legislation. But before any of this, the founders need to be clear about the route leading to product launch, and they should develop a business plan that covers each stage.

A typical start-up company sets up with an early-stage demonstrator of a new medical device, hopefully protected by a patent or patent application. To start trading, the business also needs to have secured some form of initial funding to get it up and running. Its greatest asset is of course its people –their knowledge and their commitment. Progress then follows a prescribed route.

- The company develops a working prototype that is robust and presentable enough to demonstrate in a simulated setting to potential users.

- Armed with the prototype, a patent and a favourable user feedback report, the company then raises further money to develop a commercial version with regulatory approval and evidence of clinical efficacy.

- The company then either sells the product rights to a large company, or it raises yet more cash to fund its own commercial launch and growth. The first of these options is usually preferred if available, whilst the second can involve many years of growing a viable international network. This is why the UK has very few medium-size medtech companies.

There are risks of various types attached to each of these stages, and the dropout rate is significant. But moving a new medical idea forwards is exhilarating, especially one's own invention. There is a great sense of fulfilment to be had in watching a surgical procedure where your device is improving the clinical outcome, or in encountering a patient who has benefited from the technology you developed.

A useful way to anticipate and minimise the risks is to develop a business plan that maps out the development process and assigns resources to deal with all of the anticipated issues. But before you prepare this, you should consider who you are going to approach, because the plan needs to be customised to suit the audience.

17.3   Finding the funds

There are various possible sources of funds for a start-up medtech company. A good place to start is grants. The UK Department of Business, Energy and Industrial Strategy and the Department of Health have schemes for funding innovation in small companies which are accessible from their websites. Until the Brexit referendum, funds were also available from the EU Horizon 2020 programme: UK access to this is now unclear. In general, it is easier to win grant money for research than for development, so grants are more likely to be useful at the start of your journey. The good news about a grant is that it is a genuine gift – you do not have to repay it or give away equity in the business. The bad news is that the application processes tend to be competitive and bureaucratic, and there is a considerable amount of report writing required once the grant is awarded. The expenditure rules also need to be tightly followed to avoid expensive disputes over eligibility.

Another obvious port of call is your university. Most universities have exploitation companies of their own, or links with a preferred venture capital company. Often they also have incubator facilities where you can be accommodated with access to lab space, specialist facilities and administrative services at a subsidised cost.

But if you aren't able to get reasonable terms from your alma mater, you can turn to the open market where there are two distinct funding sources: business angels and venture capitalists. Business angels are high net-worth (i.e. rich) individuals who are happy to invest a few thousand pounds in a venture that sounds exciting or worthwhile, and the medical field is usually popular. Often they will form groups or syndicates which can invest up to about £250 000. There are tax schemes to encourage these investments, and you need to understand the terms of the various options in order to ensure your pitch qualifies and is attractive.

Venture capital companies are a different breed, working to strict commercial objectives and typically investing £1 million and upwards in specified types of business. You should study their websites to understand the types of investment they support. The great majority of British VC companies will not invest in start-ups. Those that do often have preferred sectors or technologies. Some will only invest if you can find another venture capital company which is prepared to act as the lead investor. You should expect to kiss a lot of frogs before you find your prince, which means a lot of funding pitches before you get offered some money.

17.4   Business plan

This is not the place for a detailed guide to writing business plans: plenty of general advice can be gleaned from the internet. One helpful point is that the modern and entirely acceptable way to present a business plan is as a PowerPoint deck. It's an easy way to present information in bullet-point format with graphs and pictures in support, and it saves you and the reader from ploughing through pages of text.

A frequent wrong assumption is that seeking a small investment will be easier. In reality, the appraisal process that a VC applies is the same for an investment of £500 000 or £5 million. Ask for the largest amount you can justify based on the potential market. This will give you a buffer to cope with unforeseen costs, of which there will be many. More importantly, it will give your development team time to concentrate on developing the product rather than constantly being diverted to seek top-ups of funding. It really is debilitating to be tied to a drip-feed agreement in which small increments of funds are released when specific targets are achieved. Quite often the targets are irrelevant to the development needs, and precious resource is diverted to pursuing an unneeded goal.

When drawing up a plan, there are issues specific to medical devices that need to be covered. Professional investors are sophisticated operators who understand the market and the technologies and generally they do not find medtech start-ups to be an attractive prospect. Medical devices have a long gestation period because of the regulatory process; there are significant barriers that make it hard for a start-up company to generate sales; a successful device risks being copied by a large company; there is a high risk that the venture will fail for technical, clinical or market reasons; and even if it succeeds, the investor will have to wait some years before profitability and a relatively modest financial return. Recognising and addressing these points in a business plan will give comfort to potential funders that you understand their concerns and have taken steps to de-risk them. So here are some pointers...

17.5 Regulatory approval

The regulatory environment is set by the Food and Drugs Administration (FDA) in the United States and the Medicines and Healthcare products Regulatory Agency (MHRA) in the UK which regulates under the European Medical Devices Directives (though this may change). The American and European markets together represent about 80% of the global medical device market and it is sensible to gain clearance for selling in both these territories. There are moves to harmonise the regulations, but today the two jurisdictions have to be approached separately. Before you start your formal development, make sure that you have understood the two approval processes, and design your development plan to comply with them. Depending on your financial resources when you start, it may be possible to obtain a CE mark and/or FDA clearance for your device before you approach a funder. This will greatly increase the attractiveness of your proposal as it removes a significant risk and reduces the time to market. So try to plan a development path that results in a pre-production version relatively early on, which can be used as the basis for your regulatory application.

Both the FDA and CE mark schemes operate a classification system for devices which relate to their perceived riskiness, ranging from a low-risk Class 1 device, such as a hospital bed, to a higher risk Class 3 device such as an instrument for neurosurgery. Once you have determined which class your device falls into, you will be directed to the designated approval regime which reflects the risk.

Whatever the classification, approval is much simpler if you can identify a precursor product – an existing device from which you can show that yours is a logical but novel derivative. This saves you literally years of clinical trials, because you can rely on the results of the precursor to demonstrate that the scientific and clinical basis of your device is in principle safe and effective: your task is to demonstrate that the novel features you have introduced do not compromise this.

If you cannot identify a precursor, then you have to show that your device is at least as safe and efficacious as the current method. Make sure you choose criteria that will not take many years to prove – if your claim rests on showing that your device lasts longer than the 10-year life of the existing method, you will be waiting a decade for approval.

Many companies hire expensive professional support to prepare their FDA and MHRA submissions, but this is not usually necessary. If you approach the MHRA in person at the start of the process, they will helpfully guide you as to exactly what evidence and documentation they want to see, and you will be able to form a view as to whether you need help in constructing this.

To gain regulatory approval for a medical device, both the development and the manufacturing must happen in a quality management environment that documents traceability and testing. The document you need is international standard ISO13485:2016 (www.iso.org/iso/catalogue_detail?csnumber=59752), which describes the normative environment for all products in the medtech sector. A company that is accredited to this standard is accepted as meeting many of the essential requirements of the regulatory process; without it, the approvals regime is very much more demanding. Even a micro company with two or three employees will find the regulatory process easier if it is accredited to ISO13485, and the fact you have achieved it will appeal to a potential investor as it demonstrates a level of professionalism. But if you really do not have the appetite for setting up the processes involved, an alternative is to work with a contract manufacturer who has this accreditation. This manufacturer will, for a fee, be able to help with testing and documentation for the regulatory process too.

17.6   Intellectual property (IP): patents and design rights

You are most unlikely to get funding for developing a device unless you have patent protection. Without it there is nothing to stop a well-endowed competitor from copying your idea and flooding the market. If your idea came from university-based research, then the IP rights will need to be negotiated with the university authorities and will depend on your contract with them. The university may agree to cover the cost of filing a patent in which you are the named inventor, but it will then expect a considerable share of any royalties resulting. Even if the university declines to file a patent, this does not mean that it has relinquished its rights. It is best to seek advice from a patent attorney or lawyer. But do approach the negotiations in the spirit of seeking a win-win outcome – if both parties stand to benefit, they are more likely to collaborate positively.

The key thing to avoid is publication. You will not be able to gain patent protection if you have previously disclosed your device. Disclosure includes obvious publication in meetings, journals or internet postings, but also discussions or written exchanges of information with companies and private individuals. Even a casual conversation in a pub can be counted as a 'disclosure'. The best advice is to file a patent application at the earliest opportunity, and to keep the development confidential until then. If you need to share the patentable details, for example with an investor, you can avoid a complaint of 'prior publication' if you get the other party to sign a Non Disclosure Agreement, and ensure that all documents are marked Confidential.

Filing a patent application in the UK is simple, online and free of charge if you do it yourself, but you would be well advised to employ a patent attorney to write the specification for you: the wrong words or a missing phrase can destroy the validity of a patent. The patent attorney's charge is likely to be between £1000 and £2000. The day you file is your 'priority date' and your invention is protected from that time, though you will still be advised to be discrete.

A patent application remains unpublished for 12 months, and you can withdraw it at any time during this period. If you are still some way from launching a product and you are reasonably certain that no-one else has filed a competing patent, you can withdraw your application in month 11 and resubmit it without charge a few days later. You will have lost your original priority date, but gained another year of secrecy while you continue developing.

Twelve months after filing, you will need to pay a fee and the UK patent application will then be published, and you will need to decide which other countries to file in. There is a time window in which you can file applications in those countries claiming the UK priority date. This is the point where costs begin to mount up, and it pays to be prudent in selecting your market. It's entirely possible to obtain protection in Brazil, for example, but in practice if someone started to infringe your patent there how would you know and what could you do about it? If you file patents in the USA and the big four in Europe – Germany, France, UK and Italy – you will have covered half the world market and that may well be enough.

17.7   Clinical champion

Especially in the UK, but also more generally, sales of medical devices depend on reputation and association. To launch a new device, the tried and tested method is to find a clinical champion: a senior clinician who is respected in their field and who is prepared to work with you to perfect your device, publish the results and act as a reference site. Often the first question you will be asked when you launch your device at an international exhibition is 'who is already using this?' If you can respond with one or more well-known names, you are on the way to securing a sale. There is rarely any financial benefit for the clinical champion in this arrangement, as medical ethics will prevent them from being rewarded. What they will gain is recognition by publishing 'their' discovery and becoming the name that is associated with this new technique. Be prepared to recognise this is a joint venture and allow them to claim credit: you will both benefit eventually.

17.8 Getting it together

On a purely psychological level, starting and running a new business can be a lonely calling. It's more fun and more robust if there are two or three of you who can share ideas, debate options and stand together when required. This is also a good way to start: an afternoon together in a quiet room with a flip chart sharing ideas and ideals. If your enthusiasm rises with time, then you can be encouraged to go on to the next step – do some sums, work out what you need, prepare a draft PowerPoint show and take a deep breath. Most medical devices started just like this.

18 Innovation and knowledge transfer

Michael A. Smith

18.1 Introduction

The subjects of this chapter, 'innovation' and 'knowledge transfer', have as varied a number of definitions as you are ever likely to find, indeed probably as many as there are people proselytising about them. In this chapter, I shall explain the working definitions that I have developed over the years and found helpful, particularly in relation to other terminology that is in general use. One factor which makes discussions around this subject confusing is the use of the word 'innovation' as a panacea to solve the economic woes of an organisation or geographic region. Such use of the word innovation, indeed overuse would be a more accurate description, can engender confusion and cynicism in people not working in the area, thus making the job of innovation practitioners more difficult.

Furthermore, the overuse of the term innovation, increases everyone's expectations, resulting in disappointment and hence the cynicism. A successful practitioner of innovation and/or knowledge transfer will always be spending a considerable amount of their time managing the expectations of others. The most challenging expectation to manage is that of the financial impact, not only in terms of the amount of money that can be saved, but also the speed that such savings will appear.

Innovation is clearly about something new, but it is also important to consider that a consequence of the innovation is that it may require new ways of working and, most importantly, new ways of thinking. It will be necessary to understand these to maximise the speed and effectiveness of the development and acceptance of any new product. It is unlikely that the existing skills set in any organisation would contain the breadth of knowledge and expertise to consider and explore the range of new opportunities.

With innovation, I frequently use the phrase 'it's never too soon until it's too late'. A mistake often made is to wait until an innovative product is a long way through the development process before accessing external expertise, and then being very prescriptive about what input is required. I would suggest that it is important to be open-minded about external support and explore such involvement at the start of the innovation process. Only then can you be assured that the appropriate skills and expertise are available to supplement those that are in-house, leading to a more innovative, more rapid, and more market focused product.

The final point I wish to make in my introduction is to consider why there is value in considering innovation and knowledge transfer together. Throughout the history of our professions, physicists and engineers have had an interest in the development and implementation of new ideas, and have engaged in activities that would fall into the categories of research, or R&D, subjects which are covered elsewhere in this book. This chapter takes a wider perspective and uses the opportunity to introduce some different approaches.

18.2 The innovation process

Innovation is implicitly concerned with the identification, development, and implementation of something new. Ideally, and certainly in our professions, one would wish an innovation to result in societal or financial dividends, and as a consequence, have a measureable impact. The full innovation process therefore, generally has to involve a wide range of potentially interested parties and stakeholders if it is to be successful. This in turn requires the interchange and transfer of knowledge to maximise the speed and effectiveness of the innovation process and the maximisation of impact. Research is often a component of the innovation process and knowledge transfer will often form a component of research activity. However, I would argue that it is probably less of a prerequisite for successful research than it is for successful innovation.

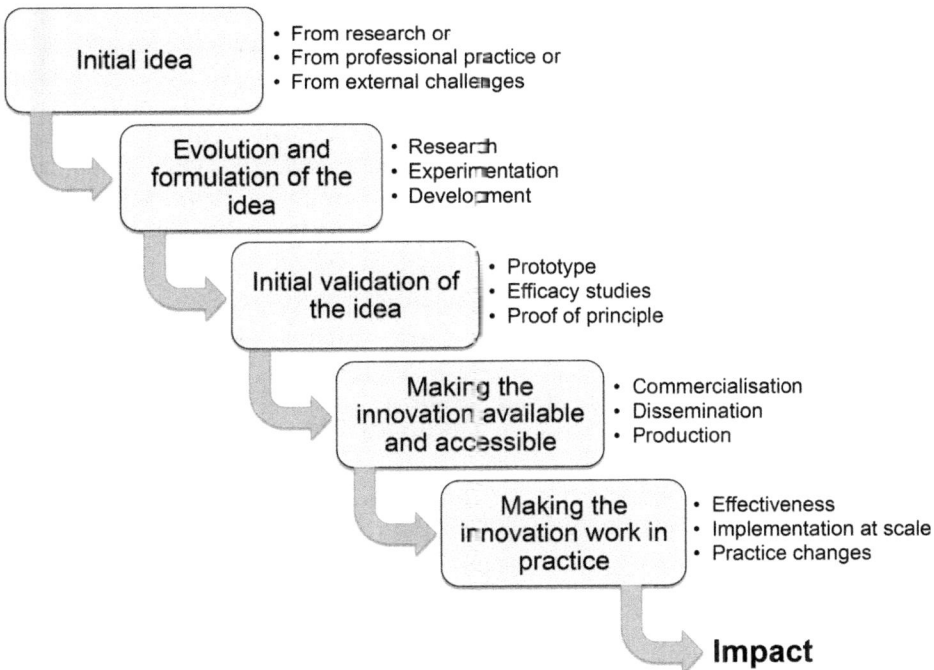

Figure 18.1 Stages of the innovation process

One model that I use to describe the innovation process, which is relevant to physics and engineering applied to medicine, is shown in Figure 18.1. It describes the innovation process from the initial idea through to the position where the impact can be measured. The second and third stages are relatively straightforward; in the case of innovations in medicine and health, it is the fourth and fifth stages that pose the greatest difficulty, both to individual medical and health practitioners and to the NHS as a whole, and each of these stages could be the subject of a complete chapter by themselves. It is these stages that are assuming increased importance as greater emphasis is placed on impact, including the research excellence framework (REF) in universities.

The model in Figure 18.1 is essentially linear, with sequential stages. In practice, there would be some feedback between some subsequent stages, in particular experience during the third stage will inform and influence the second stage, and thus influence the complete process. Similarly, the final stage can feedback and influence the penultimate stage. The value in considering the innovation process in this way is firstly to ensure that you have considered all aspects of the process, and secondly because funding agencies and investment organisations often require an explanation of the different sequential stages of the innovation process.

However, much has also been written about innovation being a non-linear process and various models exist which are helpful to consider, particularly if you have resources to pursue an innovation and are not required to justify funding before the innovation process. Alternative more generalised models, including those concerned with technology push or market pull are reviewed elsewhere (Manley, 2003). For those with a wider interest in the innovation process at a governmental level, the often quoted 'triple helix model' (Etzkowitz and Leydesdorff, 2000) is important. Essentially it articulates the importance of a coordinated approach involving government, the private sector, and universities to underpin an innovation-based knowledge economy.

Whilst we are considering the innovation process, I believe it is important to highlight the value of the 'design thinking model', which I think can be relevant for innovation development in medical physics and engineering. This approach is used widely for product development and provides a slightly different insight into components of the innovation process. The model is shown in Figure 18.2 and is focused around the development and testing of prototype designs of an innovation. It considers 'design' in the widest sense and can be applied to processes as well as products.

The model emphasises the need to understand what people need and then to test that what you may be creating matches those requirements. It is about creating a mock-up and getting feedback. An adage which you may hear associated with the design thinking model is 'fail fast, fail cheaply'. The model seeks to ensure that an innovation is both fit for purpose and is produced quickly and cost effectively.

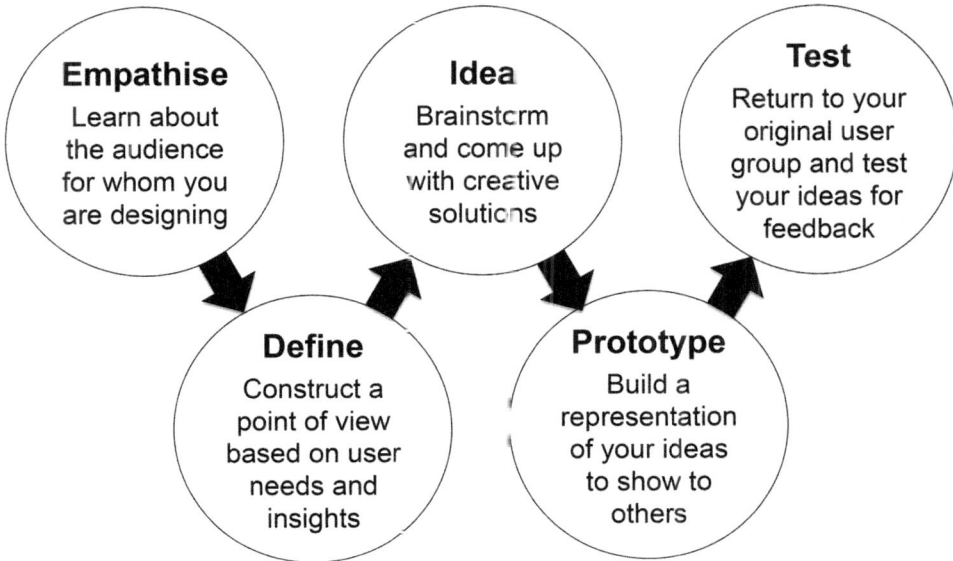

Figure 18.2 Design thinking model

18.3 Challenges and pitfalls

18.3.1 A solution looking for a problem

Often a new idea is identified with only a limited consideration or knowledge of the potential applications. In such a situation, it is easy to fall into the all too common trap of developing a solution, and then looking for a problem. To avoid this, the potential applications of any new idea should be considered in parallel with the development phase, as opposed to much later, when it may be too late. It also involves taking into account the views of customers and clients from the onset.

18.3.2 Don't leave the consideration of impact too late

The final application is often thought of as the end point of the innovation process. However, I would argue that it is crucial to give consideration to the ultimate impact of the innovation as early as possible during the innovation process. For an innovation to have an impact, there may be a requirement of behavioural change, either at an individual or at an organisational level. The identification of potential barriers and subsequent mechanisms to overcome them then becomes important, and the earlier they are known, the better.

18.3.3 Innovation or business development

It is important to distinguish innovation from what I would call 'business development', which can be associated with incremental changes. As with any sector, many opportunities exist to develop improved, or more cost-effective versions, of existing products. Such developments may have some novel features but the level of innovation may be relatively low. The route to market for an improved product is relatively clear in that healthcare professionals, who are knowledgeable and have relevant training, would be aware of its application and use, and organisations would have the experience of commissioning or purchasing the product.

18.3.4 Disruptive technology

A genuinely innovative new product faces much greater difficulties both in being accepted by professionals and getting to market. This is because, if it is doing something new, healthcare professionals may not have the necessary experience to evaluate or use it, or indeed the new product may make their existing skills and experience redundant. Furthermore, if a product has never been on the market, it will not be part of an existing care pathway, therefore new funding will have to be identified for it to be commissioned; this is particularly important if the intention is to sell to the NHS. Such innovations can be described as 'disruptive technology'. Not only may they be disrupting the existing financial and budgeting process, but they may also be disrupting how clinical care is provided and also the training and skills required by medical and health professionals. Acceptance of a technological innovation in the medical and healthcare sector requires much more than its purchase; it is also likely to require behavioural change, and it is this which often prevents newly developed innovations from being taken up by the health sector.

18.3.5 Accessing the required knowledge base

The involvement of potential users or clients in the innovation process, though important, can create difficulties if the engagement process is not facilitated properly. Users, quite understandably, will tend to use their previous experience as their frame of reference, which can restrict their initial response as to what they may want from the application of the innovation. They may not initially offer sufficient insight to the potential applications if the innovation differs from their previous experience. Translating their experience and insight into a more sophisticated analysis of needs becomes an important and often complex process; the involvement of external experienced facilitators can be enormously valuable.

There is sometimes a tendency to consider the ideal solution to a problem and therefore specify what a range of people may want, rather than what is actually needed to improve clinical or cost effectiveness. This can lead to over-specifying the requirements of an innovation, with consequential cost implications. Also, whilst medical and healthcare professionals may have a detailed understanding of elements that may affect clinical effectiveness, they may not be involved in purchasing decisions, or understand the wider economic consequences.

An elegant phrase that is very helpful when considering innovation in the medical and health sector is 'non-obvious unmet need'. The emphasis being on 'need' and not 'want', because it is the former that will impact on clinical and/or cost effectiveness. If it is an unmet need, then it should improve clinical effectiveness or health status, and therefore there should be a market, and if it's non-obvious, then it is unlikely to have been thought of, or tried before. A considerable amount of time is sometimes wasted exploring and developing innovations which provide obvious solutions to unmet wishes.

18.3.6 Route to market

The final challenge and pitfall is that of understanding the route to market for your innovation. The multiplicity of the possible routes to market makes the medical and healthcare market a complex challenge when commercialising a new innovation. Understanding the oft-times complex decision processes behind purchasing decisions also becomes central to success. The purchasers of medical and healthcare products and services include: (i) individual consumers, (ii) carers of individuals with medical and healthcare needs, (iii) private medical and healthcare providers, ranging from individual professionals to hospitals, and (iv) very large and complex public sector healthcare systems. Understanding the likely route to market will influence decisions about whether to form a company or licence the innovations to an existing company.

18.3.7 Product performance

Since the 1990s, the importance of 'evidence-based medicine' has become central in the healthcare systems of many countries and we are used to the concept of evidence when thinking about research. However, good quality evidence is equally important during the innovation process, particularly in medicine and health. Any statements about the impact of an innovation need to be backed up by data and evidence. Considerations of future impact are best addressed in association with both the invention and application stages, enabling measurements of performance and effectiveness to be considered and implemented sooner rather than later.

There are a number of key terms, which are important to understand, and which are as relevant for innovation as they are for research.

    i.   Evaluation – does the innovation perform as expected and is it safe and reliable?

    ii.   Efficacy – how does it perform on patients in an idealised situation, i.e. on patients with extreme examples of a clinical disorder, in a specialised unit with skilled and motivated staff?

iii. Clinical effectiveness – how does it perform in practice in a typical clinical/hospital environment?

iv. Cost effectiveness – what are the overall economic consequences of implementing this innovation?

It is not uncommon for an innovation to be developed followed by attempts made to implement it, without providing adequate independent evidence of any of the above criteria. To obtain the evidence, it is usually necessary to undertake tests or trials in order to obtain the necessary data and analyse it objectively. Too often, the analysis of performance is undertaken by those with a vested interest in the innovation, resulting in a conflict of interest and potential scepticism about any findings. Independent validation of an innovation is an important component of a successful innovation process.

18.3.8 Economics not just costs

A simple analysis of costs is unlikely to provide the necessary information to either encourage clinical implementation or change previous purchasing habits, if something innovative is being introduced. To change behaviour, it is usually necessary to provide compelling evidence about the impact on patient care and overall economic benefit to the organisation purchasing the product. A consideration of the health economics would also include a rigorous analysis of any potential savings, including the likelihood of them being realised.

18.4 The importance of knowledge transfer

During the innovation process, it is highly likely that external input into components of the process, i.e. knowledge transfer, will be necessary. This provides additional knowledge, expertise, and perspectives, as well as some critical challenge to 'stress-test' the different aspects of the innovation process.

18.4.1 Accessing expertise

Many organisations offer specialist support and expertise for the innovation process in the medical and health sector. Early engagement with them can be valuable, as using their general knowledge of the innovation process, coupled with sector specific expertise, enables them to provide cost-effective advice on the use of highly specialised and expensive services.

18.4.2 Knowledge as a catalyst for innovation

Ideas behind an invention are often stimulated by something which acts as a catalyst, enabling other components of the invention to work and create something innovative. Such catalysts can be new science, new technology, a new process, an implementation mechanism, or a behavioural change leading to increased acceptance of a product.

An example is X-ray computed tomography in medicine. Though CT burst on the medical scene in the 1970s, the mathematics behind the method was developed before 1920, and the X-ray technology and the electronic control systems were from the 1950s. The catalyst was the availability of 'mini computers', which enabled the calculations to be done and the image displayed in an acceptable time. The key components of CT were essentially in place, waiting for the technological catalyst, which would be the final component to enable the innovation.

18.4.3 Partnership or transactional relationship

Often during the innovation process, you may need to access knowledge from elsewhere. Simply 'buying in' what is thought to be the missing expertise is often not sufficient, because that presupposes an accurate prior knowledge of what is required. Quite simply, in a transactional relationship, you may commission someone to do something which turns out not to be what you eventually need. Establishing relationships with potential partners, based on trust rather than simple commercial transactions, can therefore be cost-effective in the long term.

18.4.4 Co-creation of innovation

The concept of 'co-creation' can be helpful to highlight the value of jointly created innovation which is based on shared ownership of the process, based around trust and mutual respect for different perspectives. As with many ideas, it is often helpful to consider what co-creation isn't; it is not simply market research, stakeholder engagement, accessing skills, or a sophisticated form of consultation.

In a co-creation environment, you would seek out advice and expertise to develop your innovation virtually from the outset, continuing to work in partnership throughout the development process, ensuring effective creative and critical support. When necessary, the co-creation relationship can be formalised in a special purpose vehicle linking two or more organisations, such as a joint venture. Furthermore, it can also provide a helpful perspective into the merits of 'open innovation'; indeed some form of co-creation may be a prerequisite for the success of an open innovation approach.

Figure 18.3 illustrates how a transactional arrangement at the bottom can gradually progress as the partners' understanding of each other's perspective increases. The nature of the relationship evolves as trust increases and a productive partnership results. The example is of a group undertaking innovation in a higher education institution (HEI) as the initial provider, working with a commercial company, which was the purchaser.

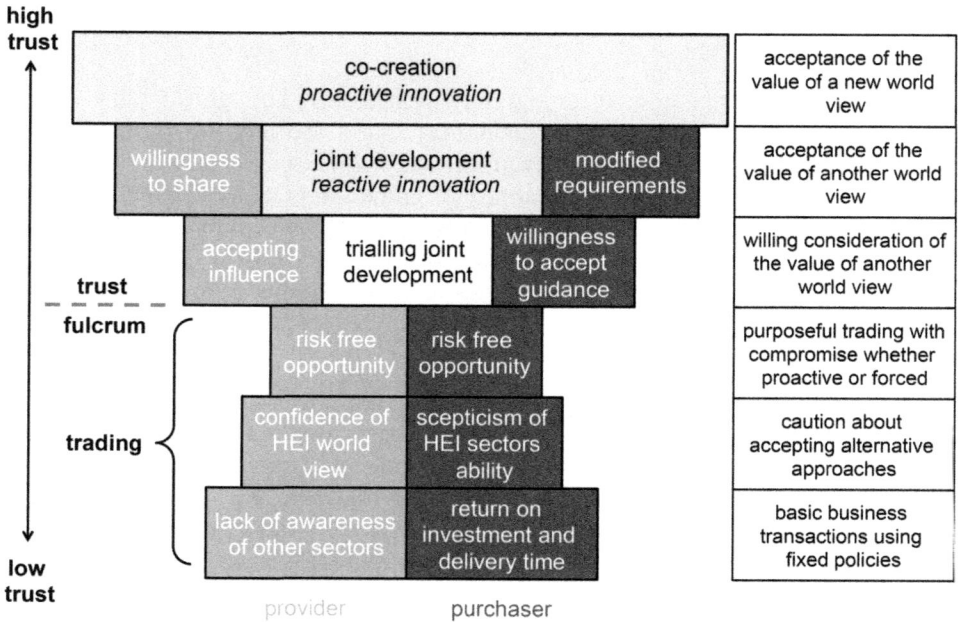

Figure 18.3 Co-creation of innovation

18.5   Innovation hubs

In 2003, the NHS created a number of 'Innovation Hubs' to identify and commercialise innovations emerging from the NHS. An example of one such organisation is Medipex (www.medipex.co.uk) which was set up to support the development of medical and health related innovations emerging from the NHS, from either NHS or university staff. They link innovative ideas emerging from the NHS with potential commercial manufacturers, allowing them to expand their product base. They also provide competitor analysis of associated products and also valuable, up-to-date knowledge and information about the current processes and systems of the UK National Health Service, which would influence purchasing decisions.

18.6   Recommendations

- When undertaking innovation, access advice from a range of medical and health professionals, innovation experts and users/carers. Do not rely on a single 'expert' in the development of an innovation to resolve an identified problem. Utilise workshops and scenario planning to capture knowledge from your partners and imagine new ways of working.

- Understand or seek advice about the health system with which you may be engaging and the financial dynamics that influence commissioning, implementation, and purchasing decisions.

- Acquire the necessary objective evidence about your innovation to support its acceptance by different healthcare systems.

- Work in partnership with organisations that can facilitate your engagement with the medical and healthcare sector and also potentially support access to funding from the sector. Consider the value of adopting a co-creation approach to innovation development.

- If you are developing new technology, recognise that behavioural and organisational change may be as important as the introduction of the new technology itself.

- When undertaking innovation development, spend time managing expectations; remember the adage 'under-promise and over-deliver'.

References

Etzkowitz, H. and Leydesdorff, L. (2000) The dynamics of innovation: from National Systems and "Mode 2" to a Triple Helix of university–industry–government relations. *Research Policy*, Vol. 29, Issue 2, 109–123.

Manley, K. (2003) Frameworks for understanding interactive innovation processes. *International Journal of Entrepreneurship and Innovation*, Vol. 4, Issue 1, 25–36.

Section 4

Teaching and learning

19 Modern educational approaches to teaching physics and engineering in medicine

Jamie Harle

19.1 Introduction

The first dedicated radiation physics course in the world was founded in 1958 as a rapid response to the Windscale nuclear plant fire of the preceding year. This was the worst global nuclear plant accident at that point in time, estimated subsequently to have led to 240 radiation-induced cancers. The inadequate medical physics expertise within the UK community at the outbreak of crisis, and further lack of skilled manpower to coordinate effective response, led two professors to establish the course. Founded by John Eric Roberts and Joseph Rotblat as the 'University of London MSc in Radiation Physics' it was, by all surviving accounts, a lecture-based tour-de-force of instructional delivery, thick with theoretical atomic physics content and lengthy descriptions of experimental apparatus. By the end of a day's blackboard writings, auditorium air was said to be almost unbreathable from chalk dust, matching the density of student note taking from the day's relentless instruction.

Professor Roberts went on to found the top scientific journal, *Physics in Medicine and Biology*, while Professor Sir Joseph Rotblat, as he became, won the 1995 Nobel Peace Prize for his life's dedication to publicising the risks to humanity from nuclear weapons. While much of their taught content in atomic physics remains valid today, the educational methods by which instruction is delivered by academic staff, as well as the expectations of the student audience in receiving appropriately designed and structured educational content, have vastly changed. This consideration of the theory and practice of education, and how best to teach a topic in a modern environment of technology-enabled teaching spaces with electronic resources, forms a discipline known as 'pedagogy'. This chapter looks at the pedagogy of the teaching of physics and engineering applied to medicine.

This London radiation physics course continues to run as a modern masters programme with a very different student demographic to that of 1958, now with a cohort including international students from all continents of the world, who learn through both synchronous campus attendance at lectures and tutorials and online asynchronous study of the taught materials at their own pace of learning. Run by the chapter author, its teaching is further arranged into three streams for physics, engineering and computing, to reflect these considerable sub-disciplines. Teaching on such streams can often be informed by current research work and uses a modern virtual learning environment to augment student studies with additional online resources and activities. Assessment is achieved by examination, coursework and project work, a small part through group assignments in collaborative groups, with clear learning outcomes written at the programme outset that define, for the student's benefit, what knowledge, skills and competencies they are expected to develop through their studies.

Student assessment is always constructively aligned: that is, it is matched against the programme learning outcomes to ensure that judgement of student performance is against a predetermined content of topics at a defined level of performance in demonstrating or applying that knowledge. This reduces student fears that examination material will be unfamiliar or at an unexpected level of difficulty, a common worry for international or off-campus student cohorts, where the social norms of the UK higher education system are poorly understood by students. Moreover, the learning syllabus is tailored to meet national accreditation standards which serve to ensure that students develop expertise and skills that meet workplace expectations, thus maximising student employability (UK Government Department for Business, Innovation and Skills, 2016). Teaching is also varied beyond simple 'passive' instructional lecturing in auditoria to include contemporary approaches that require student engagement and contribution to the learning experience, such as 'flipped learning' and other 'active' approaches discussed later in this chapter. Such modern pedagogy has transformed the possibilities for teaching physics and engineering in medicine beyond the pioneering and visionary first programme established by Roberts and Rotblat.

19.2   Undergraduate and postgraduate education

University courses involving medical applications of physics or engineering are delivered at both undergraduate and postgraduate level in UK universities. An undergraduate physics or engineering degree will often include a medical/biomedical specialisation in their degree title if the student chooses a pathway of taught modules that includes modules in medical applications in the latter years of study. At postgraduate level, there are around a dozen UK universities delivering medical physics masters programmes, specialising purely in medical application of physics for the entire programme. There are a smaller, but growing, complement of postgraduate taught courses for biomedical or clinical engineering.

It is important to differentiate clearly between the capabilities of undergraduates and postgraduates. The entry standard for professional recognition of an individual by a learned society, such as chartered scientist/engineer status or full membership of a professional body, is often set at educational attainment of a postgraduate degree with an appropriate period of structured and relevant supervised workplace training, commonly defined at 3 years. The rationale for this postgraduate educational requirement is explored next.

19.3 Quality of learning

SOLO taxonomy is an educational theory developed by Biggs and Collis (1982). It is a classification system which distinguishes between different levels of student performance when producing academic work or answering questions, and provides insight into ways that student assessment can be designed to differentiate between these levels, or qualities, of student performance. This taxonomy is presented as four levels of learning achievement, as shown in Figure 19.1. A fifth level, which is the lowest level and defines incompetence in a subject area, is omitted. From left to right, the figure shows increasing complexity in student thought processes when accurately working with concepts in their field of study. Each level is associated with a lesser or greater advancement in the synthesis of acquired knowledge and the ability of the student to link this to prior knowledge, as discussed next.

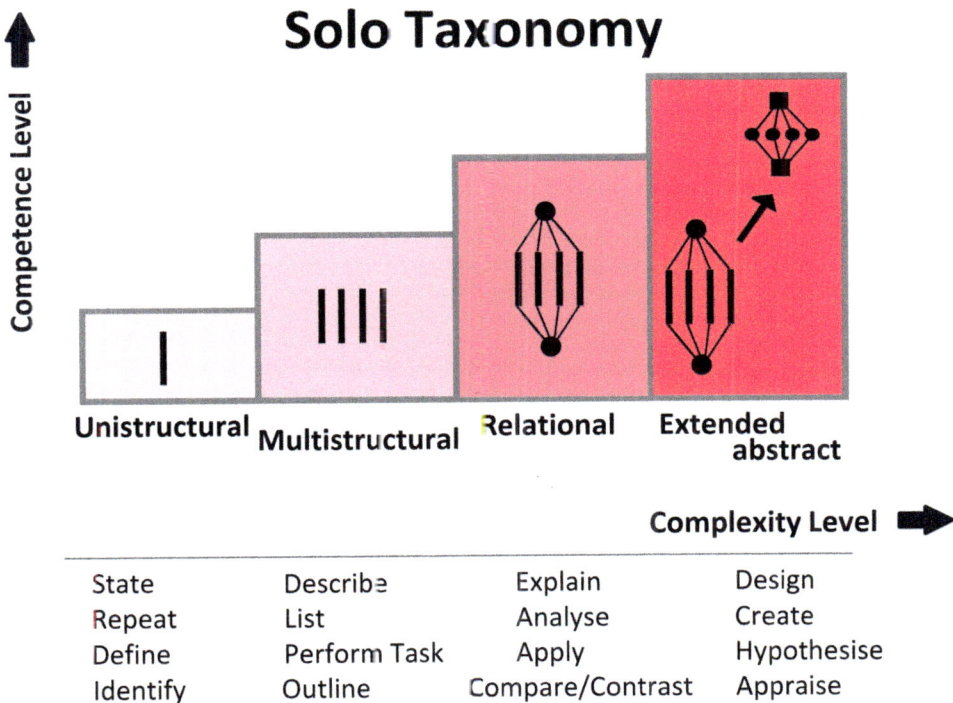

State	Describe	Explain	Design
Repeat	List	Analyse	Create
Define	Perform Task	Apply	Hypothesise
Identify	Outline	Compare/Contrast	Appraise

Figure 19.1 Using SOLO taxonomy to characterise levels of complexity in student performance and learning. The verbs underneath the four taxonomies are typically attributed to each level of cognitive thought

- *Unistructural problems* are those that exploit simple and direct connections, such as recall knowledge of facts or simple mathematical relationships. This recall by students concerns one relevant aspect of the subject area (an example: 'determine the Larmor frequency in an MRI scanner with the following static field strength').

- *Multistructural problems* are those which demand the formation of a number of connections between different learned concepts, but in which overall meta-connections within the system being studied are missed, leading to an inability to propose whole system or real-world solutions (an example: 'in X-ray imaging, discuss the implications for image quality and stochastic risk of cancer from a high and low exposure protocol'). Here, aspects of image optimisation can be discussed by the student but there remains an inability to propose an overall imaging protocol solution that considers multiple interlinked variables that govern overall imaging unit performance, such as image processing characteristics, consequences of mild thermal damage to the equipment, or quality assurance implications. Here, a student can only really support the activities of a more senior colleague who possesses such inter-relational understanding of the whole optimisation process; the student cannot act independently to autonomously propose or implement a best solution.

- *Relational problems* are those where the student can appreciate the significance of many, if not all, of the multiple concepts that influence the workings of a real-world system, and correctly explore potential solutions, each with pros and cons, and compromises and assumptions (an example: in radiotherapy, this may involve designing a protocol for adaptive radiotherapy treatment of patients that optimises clinical outcome while taking into account wider considerations for the radiotherapy service, such as quality assurance and staffing implications). At this level, students can contribute constructively to real-world scenarios featuring imperfect conditions or imprecise information. As a result, the student is now increasingly able to work autonomously, 'adding value' to projects in the workplace through his/her individual ideas and direction, particularly so with increased experience and training. Consequently, this taxonomy level forms the normal baseline requirement for recognition as an independent and competent practitioner, where professional recognition is normally attained.

- *Extended-abstract problems* involve the student exploiting past learning from relational thinking in order to generate original thought in a new field of investigation. Such work generates a hypothesis and the linking of past learning experiences, or concepts published from others, to a new scenario. Success at working sustainably to this level of cognitive thought often forms the basis of successful doctoral work, if it can be defended robustly and shown to be original and contributory to scientific literature (an example: any successful PhD thesis).

Unistructural and multistructural problems form the mainstay of undergraduate studies, while masters level is based solidly around success with the third taxonomy level of relational thinking, involving comparison, analysis and synthesis of learned concepts rather than recall of knowledge. The example verbs given in Figure 19.1 further illustrate this point, showing how assessment activities should typically vary in their complexity between undergraduate and postgraduate study, requiring differing levels of thought process from the student.

19.4   Contemporary models of learning

This emphasis on promoting higher taxonomies in student work has implications for physics and engineering courses applied to medicine. Many UK students in this field are studying while working, often as day release from clinical or industrial duties, while overseas students will be living away from home in an unfamiliar culture, or studying online using virtual learning environments. Campus-based students may also increasingly tap into online resources, such as video lectures and interactive exercises. As such, the learning experience can be increasingly individualistic with learning events or private study patterns scattered among the cohort. This is in stark contrast to the synchronous chalk-and-talk methods of the Roberts and Rotblat era, where the pace of learning was forced to match the blackboard writing and spoken dialogue of the lecturer, to be thereafter erased from historical record with the wiping of the blackboard duster.

McConnell's three models of learning in a technology-enabled environment (McConnell, 2006), summarised in Table 19.1, are good yardsticks in assessing a modern taught programme's pedagogical ethos. They describe three typical approaches to teaching in a modern university environment. Aspects of educational technology that can facilitate the more advanced approaches form the basis for discussion in the subsequent section of this chapter. In model 1, there is a one-directional transmission of information from lecturer to student, whose role is to remain passive throughout the teaching, absorbing information with the exception of occasional questioning opportunities. This was the style in the era of Roberts and Rotblat. Content is fully planned and defined by the lecturer, who is not often challenged or given quality feedback on which aspects of his or her work to improve in future. This creates a uniform learning experience for all at the detriment of opportunities to examine contentious aspects of the material and develop critical thought around the topic. Model 1 is particularly prone to occurring in physics and engineering courses where lecturing is delivered by day release or by block methods

Table 19.1 Highlighted features of McConnell's three models of technology-enabled learning

Variable	Model 1	Model 2	Model 3
Description	"transmission-dissemination"	"transmission with discussion"	"a learning community"
Role of teacher in learning & assessment	teacher regarded as the unchallenged expert & the sole assessor of student performance in the learning environment	teacher serves as a lecturer initially and then as a moderator in student activities	fully collaborative learning occurs, with teacher as the facilitator of activities, and only demonstrating authority in deadlocks, group mis-direction or final assessment for academic credit
Learning process	students passively encounter knowledge, although there is equal exposure to the taught content for all	students have equal access to lecture material but varied learning experiences in subsequent follow-up activities	students actively construct their own learning experiences in a subject area, which are diverse within the group but guided by learning outcomes to clarify expected student performance
Curriculum	determined by the teacher with no adaptive flexibility	initially determined by the teacher, with flexibility introduced through subsequent student activities. Learning outcomes required to clarify expected student performance	collectively constructed from group activities with great flexibility to explore new concepts. Natural variation of curriculum between cohorts

with a large number of lectures in an intense timeframe to regimented timetables. This is because free discussion time with lecturers can be short, or if a guest lecturer is used, there may be little to no opportunity for later consultation and student discussion. In online education, model 1 represents the simple recording and playback of videos of lecture content, without functionality for questioning the lecturer afterwards or achieving learning outcomes through formative (i.e. not for academic credit) exercises to supplement the lecture delivery such as tutor assignments or online group discussion.

In model 2, the dialogue is two-way to exploit this diversity through additional tutorial assignments or interactive sessions with the lecturer, and as a result, students are less inclined to view him or her as an unchallenged expert with the sole privilege in the group of judging student work and classroom ideas, as in model 1. A transmission-with-discussion approach also opens up student thinking to higher orders of SOLO taxonomy, where ideas can be discussed for their merits and pitfalls, and critical thinking promoted. However, this occurs with some loss of control over the delivered syllabus. Clearly, some concepts will be explored in greater depth than others. In model 2, the lecturer is ideally part lecturer and part moderator to facilitate student interaction and discussion, rather than an uninterrupted instructor. In technology-enabled education, model 2 requires supplementary formative activities, such as tutorial questions or exercises with tutor feedback to achieve the learning outcomes, above and beyond simply watching the lecture content through video.

The third model of McConnell removes any hierarchical prominence of teaching staff over the student apart from completion of the final summative (i.e. for academic credit and grading) assessment. During timetabled teaching sessions, staff serve to ensure steerage of group activity towards task completion and coverage of the learning outcomes of the assignment, but the learning journey is advanced through student effort in conducting research beforehand that they bring to sessions. Face-to-face activities serve to allow students to analyse this acquired information collectively in a structured manner in order to complete an assigned task. Problem-based learning (PBL) and scenario-based learning (SBL) are two common third model techniques where contact time is used to focus on proposing the best solution to a complex workplace or professional situation, or to developing a design or process solution to a real-world problem. This approach again pushes learning activity towards the higher complexity levels of SOLO taxonomy. Such active engagement is often deemed 'deep learning' as it promotes relational or some extended abstract thinking from students in their information gathering and discussion of their findings, in contrast to more fact-based 'surface learning' of a subject associated with less complex SOLO taxonomy coverage.

19.5 Best practice for individual teaching sessions

To this point, this chapter has discussed educational best practice from the viewpoint of the co-ordinator of teaching activities for a whole university programme or at least a substantial taught modular component. In such cases, student time for private study and reflection, as well as preparation time between teaching sessions, can be utilised in the educational design of the programme. However, there are simple, modern approaches that can be taken by the individual teacher with a minor or occasional contribution to a taught programme, which can deliver a more active and contemporary teaching experience, even with the limitation of a single teaching session.

One such approach is 'flipped learning'. This technique involves moving the instructional or knowledge-based aspects of the lesson out of the classroom and placing them onto an online virtual learning environment, most commonly as a video lecture or spoken-over presentation. A brief is then given to students to review content in their private study time before the face-to-face session. Class contact time can then be spent discussing the delivered materials or more commonly used to work through associated tutorial problems or scenarios. Thus, the teaching is flipped in the sense that the lecturer first works with students on related tutorial problems, and then uses other parts of the session to reaffirm theoretical content that is shown to be commonly misunderstood. This contrasts to a more traditional lecturing style of content first and related tutorial problems afterwards. This approach works well in physics and engineering disciplines where, particularly at masters level, there are trainee clinical/industrial practitioners in the student cohort with relevant workplace experience from whom the cohort can gain valuable insight.

The main technological enhancement that has enabled flipped learning to thrive in the university teaching environment is the development of audience response systems, colloquially known as 'clickers', in which each student is given a handset in the classroom through which they can respond to impromptu or pre-prepared questions from the teacher. This offers a more dynamic and varied teaching style to simply running through problem solutions on a whiteboard. Modern clicker systems enable each student to respond to set questions through wireless communication to the lectern PC, often in the form of answers that are true/false, multiple choice or numerical in format. Collective results can be displayed rapidly on a projector screen in histogram format, and findings used to drive classroom discussion, providing an engaging teaching style that is subtly active, varied and student-centric.

A second common approach is research- or practice-led teaching, in which a guest lecturer prepares a teaching session that overviews their research specialism or their professional role in the provision of clinical or industrial practices. This framing of academic content around local laboratory or workplace activities can be insightful for students, particularly those at postgraduate level for whom the realities of career choice and the suitable attainment of workplace skills are strong drivers of motivation. In addition, the ability to connect students to local research group activities is a current trend adopted by many UK university institutional strategies (University College London, 2016). Where local research capability is not strong or accessible for students, a variant of this approach is often adopted, known as research- or practice-informed teaching. In these methods, a selection of published work from the scientific literature or the working practices of a workplace setting (often for courses on which students can undergo placements) is presented at the onset of teaching and a scrutiny of key themes used as the basis for classroom activities and discussion in the teaching timetable.

19.6   What constitutes good pedagogical practice?

Evidence from the annual UK Higher Education Academy-Higher Education Policy Institute (HEA-HEPI) survey of university students (Neves and Hillman, 2016) shows trends across UK higher education, as well as information on how students perceive their university learning experience. This last matter is highly topical at the time of publication given the imminent roll-out of the Teaching Excellence Framework (TEF) in the UK, akin to the long-standing Research Excellence Framework (REF), as the flagship policy within the Higher Education and Research Bill passing through the UK parliament at the time of going to press. All consultation documents on this bill suggest that student satisfaction, measured from the undergraduate National Student Survey (NSS) or the Postgraduate Taught Experience Survey (PTES), will be used as a key metric in quantifying university teaching standards, along with other factors such as employability data and student completion rates.

The HEA-HEPI survey provides valuable insight into how courses can optimise student satisfaction for success in future TEF assessment. Figure 19.2 shows 2016 survey data from over 15 000 student responses, ranking students' impressions of six key characteristics of good university teaching staff (Figure 19.2a). Additionally, student exposure to each characteristic during studies is ranked and quantified (Figure 19.2b). The data shows, in contrast to the institutional strategies of many UK universities, that students rate exposure to research activity as least important, and furthermore feel they receive significant exposure to it already. Moreover, students do not feel enthused or engaged with any innovative teaching methods on offer, rating this aspect lowest for both relative importance and exposure. Attributes that score highly (over 50%) all relate to the involvement of high quality teaching staff in teaching activity, who can demonstrate up-to-date knowledge in their field, showcase modern teaching practices and evidence qualifications and training in pedagogy. Notably, the importance of clinical or industrial insight, a vital component of physics and engineering courses, is rated as only moderately important.

(a)

Rank	Student opinion on the <u>relative importance</u> of six characteristics of their university teaching staff	Rated "very important '
1	Their subject knowledge is maintained & improved regularly	58%
2	They have received training in how to teach	57%
3	Their teaching skills are maintained & improved regularly	50%
4	They have relevant industry or professional expertise	47%
5	They employ original/creative teaching methods	37%
6	They are currently active researchers in their field	26%

(b)

Rank	Student experience of <u>receiving adequate exposure</u> to each characteristic found in their teaching staff	"Demonstrated a lot"
1	They have relevant industry or professional expertise	42%
2	They are currently active researchers in their field	38%
3	Their subject knowledge is maintained & improved regularly	35%
4	They have received training in how to teach	21%
5	Their teaching skills are maintained & improved regularly	18%
6	They employ original/creative teaching methods	16%

Figure 19.2 *(a) HEA-HEPI 2016 data summarising what university students perceive as most important traits in university teaching staff. (b) Corresponding data showing the exposure to such traits among the 15 000 student respondents*

In a UK higher education sector where students are increasingly perceived as paying consumers demanding good teaching standards and employment prospects, such findings indicate to the author that a joint academic-guest lecturer approach within taught modules may be the best strategy for courses teaching physics and engineering in medicine in the future. University-based staff, with refined skills in teaching delivery and a tool bag of innovative teaching expertise to offer an active learning experience, would be tasked with delivering the broad theoretical underpinning of a subject area, which should promote student engagement, with the higher SOLO taxonomy skills that form the educational gateway for admission to professional recognition. Invited guest lecturers then add valuable contributions from the hospital or industry perspective, offering relevant workplace insight to augment this theoretical knowledge, which is a time-efficient use of their limited time for teaching commitments.

19.7   International student considerations

Consideration of educational best practice would be incomplete without briefly addressing the needs of the international student. Such students contribute to HEA-HEPI survey data, but demand additional consideration given the cultural challenges they face in successfully adapting to UK university life. Moreover, non-EU students contribute higher tuition fees, which are vital to maintaining the financial solvency of many UK courses, including many in physics and engineering disciplines.

The proportion of international students at UK higher education institutes has, surprisingly, remained quite constant over the last 60 years, varying between 10% and 12% (King *et al.*, 2013). What has varied markedly, however, is the country of student origin, with a steady growth in non-Anglo-Celtic backgrounds in STEM subjects, such as North Africa, the Middle East and East Asia, at the expense of reduced Commonwealth entry where more anglicised behaviours are derived from historical associations. In his seminal work on the learning approaches of international students, the central argument of Biggs (1999) is that variation of student performance within an international cohort arises mainly from cultural differences within the class, specifically relating to classroom behaviours developed during schooling. This factor is dominant over others associated with language, provided simple communication rules are followed in the classroom (i.e. appropriate pacing of the lessons).

In STEM subjects, international cohorts consist heavily of students from China, Malaysia and Singapore (often referred to as Confucian Heritage Culture, or CHC, students). One common trait amongst CHC students concerns how to interact with the teacher; CHC cultures deeply respect authority, and their young adults feel that challenging a senior figure is highly inappropriate, thus leading to the common impression by educators that CHC cultures learn passively without active engagement (Egege and Kutieleh, 2004). This is extenuated by a cultural trait in CHC students to behave differently in formal environments, such as a classroom, compared to informal environments, such as a university canteen. This means that CHC students will often adopt an introverted persona in a teaching environment that is not reflective of their true personality, reinforcing a perception that CHC students fail to adapt easily to situations where self-direction, creativity or critical thinking are required. The most serious implication for this apparent skills deficit, compared to UK students, concerns the ability of CHC students to therefore engage successfully in active learning environments, such as Problem Based Learning modules, where self-direction is required in completing preparatory work, and critical skills needed to evaluate and prioritise the information gathered by the cohort. Other cultures feature other challenging traits that demand lecturer consideration in course design, such as a reliance on rote (memory-based) learning, poor creativity or a tendency to disseminate personal opinion rather than evidence-based information (Biggs, 1999).

179

19.8 Summary

This chapter discusses educational approaches that promote participatory methods of student learning, in contrast to the transmission-dissemination model applied by Rotblat and Roberts in the 1950s. Such methods have the advantage of promoting a 'deeper' understanding of concepts through establishing learning experiences where students must acquire and synthesise knowledge, and thus engage more with higher SOLO taxonomies in the processing of information relating to their study topic. This in turn promotes good employability skills in physicists and engineers. However, such approaches come at the cost of reduced control by the academic designing the teaching over curriculum content, and may introduce challenges for overseas students with different cultural norms within the learning environment. Such contemporary approaches may, however, prove less popular with students, and this is an important factor in the years ahead where teaching quality and student satisfaction with their university experience look likely to be closely scrutinised.

References

Biggs, J. (1999) *Teaching for Quality Learning at University*. Open University Press.

Biggs, J.B. and Collis, K. (1982) *Evaluating the Quality of Learning: The SOLO Taxonomy*. Academic Press, New York.

Egege, S. and Kutieleh, S. (2004) Critical thinking: teaching foreign notions to foreign students. *International Education Journal*, Vol. 4, Issue 4, 75–85.

King, M., Creaser, C. and Matthews J. (2013) *A National Survey of UK HE STEM Practitioners*. Higher Education Academy.

McConnell, D. (2006) Chapter 3: The experience of e-learning groups and communities. In: *E-learning Groups and Communities of Practice*. McGraw-Hill, New York.

Neves, J and Hillman, N. (2016) *Higher Education Academy – Higher Education Policy Institute (HEA-HEPI) Student Academic Experience Survey*. Available at http://www.hepi.ac.uk/wp-content/uploads/2016/06/Student-Academic-Experience-Survey-2016.pdf

UK Government Department for Business, Innovation and Skills (2016) *The Wakeham Review of STEM Degree Provision and Graduate Employability*. Available at https://www.gov.uk/government/uploads/system/uploads/attachment_data/file/518582/ind-16-6-wakeham-review-stem-graduate-employability.pdf

University College London (2016) *Connected Curriculum: Enhancing Programmes of Study*. Available at http://www.ucl.ac.uk/teaching-learning/connected-curriculum/Enhancing_Programmes_of_Study_Sept_2016

20 Teaching-track careers in academia

Jamie Harle

20.1 Introduction

Universities in the UK require recognition by the UK Privy Council, with over 150 recognised higher education institutions (HEIs) in 2016. The status of teaching specialist staff is subject to considerable variation between HEIs, some of which prioritise research activity over teaching, while others treat those practising across any of the full spectrum of academic duties – research, teaching, academic administration and enterprise activity – with more equivalence. This chapter begins by outlining three distinct groups of HEI within the UK university sector.

- *Russell Group universities* – 24 UK institutes with strong, sustained research reputation and significant grant income across a broad spectrum of subject areas, often with historical dates of foundation and reputations that lead to high world university league table rankings. All have sizable academic departments of physics, engineering and medicine, apart from the London School of Economics. Redbrick universities are those in this subset with late 19th/early 20th century foundation.

- *Plateglass universities* – around 20 UK institutes formed in the mid-1960s from the expansion of the UK university sector. Typically, these institutes have developed niche areas of research excellence that compete against Russell Group researchers for funding in specialised fields, but lack a broad academic coverage across all disciplines. This can impact on their ability to support large multi-disciplinary research or teaching activities. Some may not have academic departments of physics, engineering or medicine.

- *Post-92 universities* – around 35 institutes formed by the UK Further and Higher Education Act (1992) which empowered Privy Council recognition of existing technical or central institutions. A relatively small number have physics or engineering departments (e.g. Hertfordshire for physics, Central Lancashire for engineering), with no medical schools. Areas of exceptional research output that compete with other groupings of universities are found, often concentrated around a small number of key academic staff in one department.

Some institutes fall between the gaps in this analysis: The Open University and St George's London, for example, both have unique foundations and remits, but are active in fields associated with this book. Other notable universities outside the invited Russell Group have a long track record of excellence in specific fields, for example, the University of Dundee (all such outliers considered to be plateglass in the chapter analysis). There are also newly recognised private and ex-further education institutes, although none are currently offering full physics or engineering programmes or undertaking high-profile research (these institutions are considered post-92 in this chapter). Successful approaches to developing a teaching track career will vary considerably depending upon the educational environment in which one is working, and thus these three broad university categories will be referred to throughout the chapter.

Typically, entry level teaching positions can be more attainable at post-92 universities, which are normally more willing to develop inexperienced candidates who show potential, than at Russell or plateglass institutes, where job specifications for teaching posts are more established and applications highly competitive. At Russell or plateglass universities, candidates for an academic teaching role will normally be expected to demonstrate previous higher education experience, sometimes in leading or establishing modules or programmes, a PhD or equivalent academic qualification, some relevant professional practice (i.e. hospital/industry work) and indeed an additional qualification in higher education. The nature of academic roles with expected teaching duties, normally categorised as lecturers or teaching fellows (in contrast to teaching assistants, who act in a non-academic support role), is discussed later.

In research careers, many young scientists or engineers progress through attainment of a multi-year funding award within a university research group, often via a fellowship or a series of competitive research grant awards. Teaching-track careers are more transitory. Developing a career and reputation in a university teaching role can often involve moving HEI at regular intervals in early career years, as openings often involve opportunities with uncertain student uptake or roles that teach generic educational content, such as introductory science or mathematics. Teaching such non-specialist material provides the benefit of being associated with large student cohorts, and therefore being regarded as a major contributor to teaching income in a university department or faculty; career stability and progression can result from this; however, it risks developing a career identity and skillset other than that of a medical physicist or biomedical/clinical engineer. This can be detrimental to ambitions to publish, network and otherwise conduct activities that develop a national or international reputation over a career.

This concern over professional identity also drives staff transition, and can lead to higher education specialists seeking new opportunities periodically in their career. One common dilemma of teaching specialists is whether to contribute to large undergraduate taught courses, often with a cohort size of a hundred or more, where one may be teaching generic theoretical or skills-based content at regular intervals, but not necessarily holding a position of responsibility. The alternative is to lead, or play a major role, in a specialist medically-orientated physics or engineering programme where cohort sizes are smaller. This second teaching role creates good opportunities to maintain and further develop physics or engineering practice, such as supervising research projects and delivering lessons relating to cutting-edge developments at the forefront of the field. This teaching duty is often more varied in workload activities than the first role, so enabling a higher education professional's skillset to continue to develop: one can benefit from opportunities that arise in holding a niche educational expertise (i.e. educational consultancy or overseas training projects that contribute to university enterprise or international work, as discussed later). However, such UK specialist university courses typically have cohort sizes of 10–40 students for both undergraduate courses (e.g. a medical applications stream of an engineering degree) and postgraduate courses (e.g. an MSc programme in medical physics). Faculty pressure to maintain, and ideally enhance, student numbers and its teaching income can be intense, especially if course numbers fall alarmingly for a single year, as is prone to occur at masters level where overseas student recruitment is volatile and influenced by geo-political factors, such as currency exchange rates, homeland security issues and changes to UK student visa requirements which can all discourage applications.

Thus, employment histories amongst experienced university teaching professionals can often extend across more than one, if not all, of these three described university groupings during a career, due to a range of issues concerning job security, skills development, professional identity and the opportunity to diversify into new work within the wider remit of the university.

20.2 The standing of teaching activity within UK universities

All 24 Russell Group universities refer explicitly to research and teaching as core activities within published mission statements (University of Birmingham, 2015). This compares to 18 out of the 24 which promote enterprise activities within their mission statement, and 11 which target the internationalisation of activities, such as addressing cross-disciplinary global challenges such as climate change. This analysis would suggest that teaching is held in equal esteem to research activity, certainly within the Russell Group.

However, few university teaching professionals would agree with this, and instead would be likely to argue that a number of barriers exist that restrict their onward progression to a permanent position, and then subsequent promotion to senior university roles in later years. This, in reality, relegates the status of teaching activity below that of research. The most dominant factor relates to the investment of university funds into each university mission activity, and the resultant future income generation that this will garner.

UK universities vary compared to their American counterparts in that few institutes receive significant alumni donations or endowments, apart from a handful of ancient institutions, such as Oxford and Cambridge, with assets from historical patronage. Thus, income must at least equal the expenditure at UK HEIs to ensure financial stability, and indeed most UK universities run a surplus of less than 4% from which capital projects must also be funded (Higher Education Statistics Agency, 2016). UK universities also differ from their European counterparts in receiving lower levels of financial support from the state for their teaching mission. This is evidenced by a gradual withdrawal of state funding to pay for student tuition, a process which began following *The Dearing Report* (The National Committee of Inquiry into Higher Education, 1997) and accelerated in 2010 (Department for Business, Innovation & Skills, 2010), albeit with the devolved UK governments enacting varied policies, such as the full payment of tuition fees for Scottish domiciled students by their devolved administration.

Higher education sector data shows that total university teaching income between 2009/10 and 2013/14 rose from £17.3 billion to £19.8 billion per annum (a 14% increase). This has occurred in a climate of considerable reform in how teaching fees are paid, with costs being mostly transferred from state to the individual through a student loans system (Higher Education Statistics Agency, 2016). However, this was against RPI inflation within the analysis period of 17%, indicating no real-term change. In contrast, research council income from competitive grant calls rose by 30% to reach £5.6 billion in 2013/14. Therefore research activity is viewed in recent times as the more government-encouraged activity, with real-term gains in funding. The second point to note is that university teaching income has remained stable in inflation terms despite widespread upheaval in funding mechanisms in recent years. This perpetuates a long-held view that teaching is a highly ubiquitous and resilient university core activity: one that is dominant in terms of monetary catchment, with over three times the total purse of research income, but one that can weather storms well and survive some neglect of capital or staffing outlay. This is important: with a limited surplus from all activities, universities see investment in research infrastructure, and attracting accompanying best expertise, as better value. In contrast, investment in teaching facilities and teaching expertise is de-prioritised.

The viewpoint is similar for individual academics, perpetuating this cultural view of the status of teaching. To the seasoned research-active professor with an additional teaching brief, the outlook can be anecdotally summarised in the following typical scenario: 100 students enter his or her compulsory first year module each year and the same number, minus a few losses to student drop-out, are seen at the corresponding graduation event for finalists in later years. From the professor's viewpoint, the demographic of the cohort may have changed over time (more overseas, more self-funded students), but the teaching activity he or she undertakes seems less sensitive to sector change than corresponding research activities. In the laboratory, changes to research council priorities for support can decimate or considerably expand his or her research group. Similarly, investment in new senior research expertise or significant laboratory equipment can generate better grant applications and open up new avenues of publication and income. From the professor's viewpoint, investment in teaching personnel is not seen to impact much on improving the performance of the university.

Thus, the pressure on university heads of department from senior academics is predominantly to hire early-career academics with a heavy preference given to their research capability and compatibility, over demonstration of excellence in teaching performance or subject coverage. Staff with novel skill sets to complement existing research activities are seen especially favourably, in offering new avenues of progress while also making best use of existing infrastructure. Staff with fresh approaches to teaching are not normally viewed as offering workplace transformation. In short, research grant income is seen to depend heavily on the quality of the infrastructure and staff that surround a bid. In contrast, teaching income is associated with the quantity of students that are enrolling on the course, particularly from overseas destinations, and is viewed as relatively insensitive to conservative change in teaching arrangements. For these reasons, teaching specialist academics are often viewed with less value and prestige than are research specialist academics.

20.3 UK parliamentary legislation to incentivise university teaching quality

The UK university sector is a major contributor to the country's economic productivity generating £10.8 billion in export earnings (2.8% of UK Gross Domestic Product in 2014) (Higher Education Statistics Agency, 2016). The sector's attractiveness to international students results in a further £7.2 billion in total spending within the UK economy from student fees and other activity. This global demand in the UK higher education system is derived from a plethora of reasons. These include excellent institutional global research rankings, with many Russell Group and plateglass institutes in the top 50 or 250 worldwide institutes, respectively. There are robust standards of educational oversight by an established external examiner system, clear guidance to academic regulations and educational best practice by the UK Quality Assurance Agency (QAA) (Quality Assurance Agency, 2008), postgraduate employment opportunities for those with appropriate visa permissions, and a highly marketable brand for UK universities framed around historic university settings and world leading alumni. Overall, the satisfaction survey ratings of overseas students on postgraduate taught courses in UK higher education is high at 89% (Universities UK, 2016).

However, the transfer of tuition fee burden from state to individual for most UK home students has led to increased scrutiny from the UK government over the quality of this educational experience given the considerable financial outlay it now involves. Average student satisfaction among UK undergraduates is only slightly below that of overseas postgraduates, at 85%, but the 2016 HEA-HEPI survey, discussed in Chapter 19, shows that only around a third (37%) of home students now consider their course to have delivered good value for money (Neves and Hillman, 2016).

A sector-wide influence in this prioritisation has been the presence of the REF (Research Excellence Framework) with no teaching equivalent. Since its introduction in the late 1980s as a tentative audit exercise for monitoring research spending, REF has been renamed and revised approximately every 5 years. By the 2014 cycle, REF had evolved into an extensive metric-based administrative exercise that definitively scored research performance of university researchers and departments. From this scoring, individual job performance is often rated, and future institute funding allocations determined for the block grant component of the UK dual research funding mechanism. With no equivalent audit process for teaching excellence, with neither institutional financial reward for success nor penalty for failure, a growing imbalance in institutional motivation has developed.

The 2016 Higher Education and Research Bill is in the first stages of its passage through the UK parliament as this book goes to press. In terms of teaching reform, the bill is set to introduce a far more student-centric outlook to teaching practice with accountability by programme leaders for student outcomes relating to the National Student Survey (NSS) and Postgraduate Taught Experience Survey (PTES) (discussed in Chapter 19). In addition, there will also be scrutiny of student employment rates, a metric indicating those in highly skilled employment after an extended period after graduation, an emphasis on ensuring equality of opportunity for students from disadvantaged backgrounds, rates of teaching-specific qualifications held by staff, student completion rates and an overall non-metric assessment of the learning environment, including availability of career development opportunities for teaching-specialist staff. There may also be initiatives to enable more work-friendly approaches to study for part-time students via flexible, accelerated or institutional credit transfer arrangements, which would suit many medical physicists or clinical/ biomedical engineers, often studying while in employment or training.

20.4 Implications for physics and engineering disciplines applied to medicine

University departments teaching physics and engineering applied to medicine will face some specific challenges in this proposed teaching assessment framework. In many such departments, local hospital employees play a vital role in the delivery of course content associated with clinical practice, such as radiotherapy quality assurance processes and treatment planning techniques. However, it may prove difficult for such valued guest lecturers to find the time commitment to undertake the training in pedagogy that would be needed to deliver new contemporary teaching methods, linked to educational theory, that can best ensure good student survey results. Moreover, the completion of a teaching qualification in higher education without reduction in clinical duties, a likely assessment category in the new framework, may be asking too much of guest lecturers or honorary clinical staff.

The list of ideal requirements for an early-career academic in medical physics or clinical/biomedical engineering is indeed becoming too exhaustive: different interview panels may seek a PhD, a postdoctoral research record suitable for a good REF performance, hospital or industrial experience in which to understand the translation of research work to the clinic (ideally with professional registration to work directly with patients), and a formal teaching qualification with proof of teaching prowess to contribute well to the Teaching Excellence Framework (TEF). With masters and undergraduate degrees to complete before these milestones, training now comprises a minimum period, assuming one step at a time without breaks, of several years beyond a full decade. This will extend training, if all steps are required, into the phase of life when raising children and seeking stable employment for financial security become dominant factors for many. Thus, divergence of job specifications into more realistic requirements may become increasingly typical, as follows at each HEI grouping.

- *Russell Group universities*: All have medical schools and large established research groups in this field, so recruitment of new academics may be dominated by matters of individual research potential and compatibility with the ongoing research work. State registration may not be needed given that large research groups often hire research radiographers, or collaborate with medical colleagues, who oversee patient work. Teaching roles may be handed to teaching-specialist staff to ensure colleagues working on research can mostly focus on delivering good REF performance. In order to supervise the research group's various student projects, teaching staff will be expected to hold a PhD; in addition they will need a teaching qualification and will have to demonstrate good academic standing on a par with research colleagues, and thus may publish in educational journals.

- *Plateglass universities*: Increasingly with attached medical schools, often of recent foundation, and normally with ambitious plans for improved REF performance. Successful applicants are normally expected to balance teaching responsibilities with developing a research profile at the early career stage. Later career specialisation in teaching or research may occur at senior levels. Here, a young all-rounder is often sought who can contribute to both teaching and research within a moderately sized group of academics. Clinical experience is the lowest of the list of requirements, but still valuable for involvement with medical school-related research projects. Plateglass universities tend to market their good student experience heavily in publicity, necessitating teaching qualifications in new staff to evidence commitment to student learning.

- *Post-92 universities*: Without an associated medical school, and with lower institutional REF expectations, a greater emphasis is placed on the teaching abilities and qualifications of applicants at interview than in other universities. Successful applicants may be expected to contribute heavily to several programmes that straddle generic healthcare science or engineering/technology topics and their teaching may be expected to be the main focus in the appraisal of their job performance. Clinical expertise to provide practice-based insight is often the second choice factor from applicants, with research success welcome but not expected. Research ambitions may require building a research group up in isolation, a challenging task!

The workforce model that is now emerging at many Russell Group universities is the increased specialisation of academic staff into teaching- or research-dedicated roles, with a number of joint posts where both roles remain expected in the job brief. There are exceptions to any rule, of course, with some Russell Group universities instead promoting an alternative where all academic staff, including senior and research-contracted individuals, must remain active in teaching and contribute to research. It is indeed possible to name inspirational academics who sustain, through considerable ability and tireless dedication, continual excellence in both teaching and research. Nevertheless, policy change from research councils continues to favour competitive funding allocation for large scale projects ('Big Science') with multi-disciplinary, multicentre collaborations a growing proportion of the competitive funding allocation. Such large scale projects often demand full-time dedication to deliver a complex role within a large coordinated team. International impact is expected from findings, ensuring excellent REF performance. Teaching specialists are needed to absorb the educational duties of such research staff. Moreover, as teaching performance is likely to be subject to increased standards of delivery for good TEF performance, this furthers institutional motivation to allocate teaching activities, certainly in flagship programmes, to those with good pedagogical practice (see Chapter 19).

Russell Group universities are particularly under pressure to improve teaching quality: of the UK HEIs ranked in the 2016 National Student Survey, nine Russell Group universities feature in the upper quartile, as expected of their status, with another nine in the second quartile. However, three feature in the third and two are remarkably placed in the lowest quartile, with student satisfaction akin to findings from new entrant universities or ex-further education colleges. Clearly, a greater focus on teaching quality will be needed to improve student satisfaction for such institutes with a global reputation and large intakes of international students. Employment of higher education specialists is one strategy being adopted to address this.

20.5   Employment opportunities in higher education

An analysis by the author of UK academic job vacancies advertised in the month of June 2016 is now presented. Briefly, UK-wide academic post advertisements in all engineering and physics fields were collected during the full month, a key period for the recruitment of staff for the new academic year. Collection was completed from jobs advertised at the most common online portal for UK academic job vacancies (www.jobs.ac.uk). Technical or administrative roles were excluded, as were studentships and senior academic roles (e.g. Professor, Group leader). Those working in translational roles (i.e. Knowledge Transfer Partnership Associates) were included, alongside postdoctoral positions for younger researchers, such as junior research associate/assistant, and the equivalent teaching associate/assistant positions. Thereafter, the salary offered by each vacancy was mapped onto the non-clinical academic range for University College London, as a fairly standard wage structure within the UK university sector, with pay bands from 5 (teaching/research associate) to 10 (professorial scale). Band 9 is for senior lecturers or readers and the top of band 6 the minimum level for staff with a relevant PhD. Teaching Fellows have a minimum salary at the midpoint of Band 7, as do new lecturers with both teaching and research remits, albeit with an incremental increase given their dual role.

Figure 20.1a shows this breakdown in terms of job title. Few opportunities dedicated to physical sciences or engineering exist in post-92 universities, where teaching roles tend to straddle whole disciplines (i.e. introductory science for healthcare professionals). Overall, there are a large number of research fellow posts, reflecting the short-term contract nature of such positions. Teaching fellow posts are less frequently available, probably because they are normally offered as a permanent contract beyond probation, as with both categories of lecturer, which lowers staff turnover. Teaching fellow posts are also seen to be a Russell Group phenomenon, with the alternative job title of Lecturer (Teaching and Scholarship), or similar, more apparent in plateglass institutions, with a decreasing number of exceptions. Such teaching appointments normally make it clear in job descriptions that research performance is not assessed in appraisals and that REF submission is not expected, although teaching specialists may still supervise research projects that lead to publication.

A third of all posts that were traditionally allocated to dual-role lecturers are now advertised as teaching-specialist positions, a proportion that is currently consistent between Russell Group and plateglass institutes. Furthermore, analysis in Figure 20.1b shows a significant proportion of band 7–8 lecturer posts that demand performance against only teaching activity. A decade ago, both teaching and research remits would have been core roles of all lecturer staff. This implies a growing divergence of skill set in early-to-mid career academics with specialisation in either teaching or research, albeit with other staff still working across both roles. One natural extension of this analysis, with the imminent dual factors of TEF and REF, may be the expectation of academic staff to contribute with distinction to

(a) Breakdown of UK advertised posts in physics & engineering - June 2016

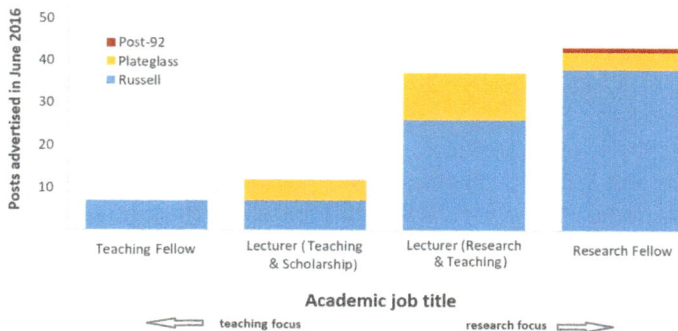

(b) Breakdown of posts by job grade (non-clinical university bands 5-8)

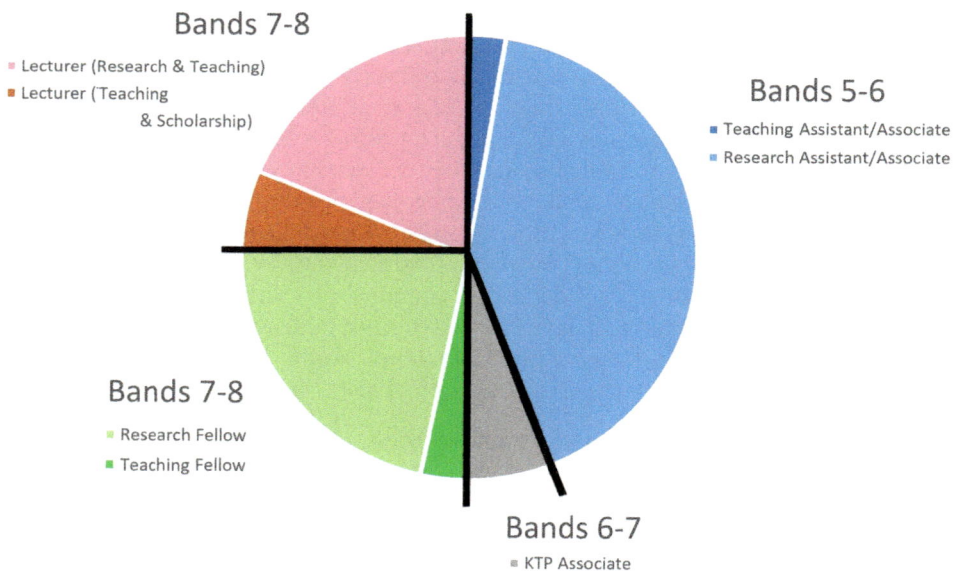

Figure 20.1 An analysis of advertised non-senior academic posts in UK university physics and engineering fields during the peak summer recruitment window.
(a) Breakdown by job title for the three university categories outlined in this chapter.
(b) Breakdown by pay scale and teaching/research focus

at least one of these two assessment exercises. This may lead to a decoupling of the academic skill set over time, with individual effort concentrated more and more at high accomplishment in only one role. Inevitably, academics may find themselves increasingly stranded in that specialism, unable to maintain or re-establish the track-record and skill set to switch between research and teaching roles.

Another possible long-term consequence of a dual TEF and REF environment may be whole departments, or even HEIs, specialising in either teaching or research as part of a strategy to maximise income by focusing on optimisation of one income source: REF block funding, or TEF income. TEF income is expected to derive from the unlocking of caps on tuition fees for universities with highly rated teaching quality (Department for Business, Innovation & Skills, 2016).

20.6   The teaching track in UK academia

The teaching-specialist track in physics and engineering is normally open to those with a PhD qualification, or with equivalent professional achievements such as chartered status or extensive industrial/clinical experience. In the field of physics and engineering applied to medicine, both may be reasonably expected for successful appointment. A teaching qualification in higher education will also be needed; this is normally expected of candidates who have previously worked in universities, or becomes a requirement of passing probation for those from industry or hospital backgrounds. At teaching fellow level (band 7), successful applicants can be considered independent practitioners with responsibility for their own output, management of a portfolio of teaching duties and control over their career development. At this lowest band, a teaching fellow will typically input to class activities with great regularity and lead modules on programmes. At band 8, they will lead programmes, or several modules in large cross-departmental teaching activities with high student numbers. They will also lead educational development of new modules and oversee activities at the more innovative or pedagogic end of a faculty's teaching portfolio. At band 9, a more strategic role such as managing a faculty portfolio of courses is expected. Table 20.1 shows how career advancement can be defined up to a professorial level (band 10). There are normally only a handful of professorial teaching specialists per HEI.

20.7   Advice for embarking on a teaching track career in physics or engineering applied to medicine

This final section offers anecdotal advice for those wishing to embark on a teaching track career in this field, either intentionally after completing a PhD qualification, or as a change in direction from a previous research, clinical or industrial career.

Table 20.1 Six common performance criteria at UK universities for teaching-specialist academics from academic bands 7 to 10

Performance criteria	Band 7 - Teaching Fellow	Band 8 - Senior Teaching Fellow	Band 9 - Principal Teaching Fellow	Band 10 - Professorial Teaching Fellow
Responsibility	Run modules and/or deliver teaching sessions in modules	Lead taught programmes and design educational delivery and/or manage important educational projects	Oversee strategic growth of a portfolio of programmes or a flagship institute-wide initiative	Sustainably lead an educational development unit or an educational research group
Qualifications	PhD (or equivalent) & FHEA	PhD (or equivalent) & SFHEA	PhD (or equivalent) & PFHEA	Awards of national significance (i.e. UK HEA NTFS winner)
Impact	Impact at departmental level	Impact at a cross-faculty level	Impact reaching national level	National/international impact
Typical scholarly output	Present work that shares best practice at internal meetings and HEI events	Present work at UK conferences to address common themes in pedagogy	Enhancing pedagogy through literature that contributes new evidence-based approaches	Generate external income to fund pedagogic research with sector-wide impact

Table 20.1 Continued

Performance criteria	Band 7 - Teaching Fellow	Band 8 - Senior Teaching Fellow	Band 9 - Principal Teaching Fellow	Band 10 - Professorial Teaching Fellow
Managerial roles	No line management roles	Line managing teaching assistants or associates	Co-ordination of activities at a faculty or service unit level	Policy input into QAA, HEA, HESA or other UK higher education bodies
Input to institutional teaching governance	Collate information on student feedback	Complete annual reviews of departmental teaching activity	Contribute to institutional audit or reviews of university practice	Lead institute strategy to teaching and learning or national educational initiatives

FHEA is fellowship of the UK Higher Education Academy, with the prefix S or P denoting senior or principal grades of fellowship. The Higher Education Academy National Teaching Follow Scheme (NTFS) (Higher Education Academy, 2016) is a highly prestigious accolade for higher education specialists, often expected of professorial staff. QAA, Quality Assurance Agency; HEA, Higher Education Academy; HESA, Higher Education Statistics Agency.

The main advice of the author is to engage with pedagogy and become increasingly comfortable with educational theory. Chapter 19 explores some basic educational theory. This is a difficult process to begin for a scientist or engineer as most educational theory is grounded in sociology or psychology disciplines. There may be a lack of quantitative data or even a complete lack of any apparent data in the literature considered seminal. The non-quantitative, discursive style of its writing can appear very unnatural at first for a trained scientist or engineer. However, contemporary educational theory is evidence-based and student survey scores do tend to increase following its implementation in programme design and delivery, re-affirming its value. A postgraduate certificate in higher education (PGCert in HE) is a common part-time route to begin learning about educational theory. Such courses in academic practice are delivered by most universities that host a dedicated centre for the advancement of teaching and learning, a university-wide staff resource at most universities normally staffed by professional educationalists. Often, there is no cost for the probationary teaching fellows to study the PGCert in HE at their institution. For those seeking to enter the higher education field from other backgrounds, the completion of this qualification before application is an excellent means to demonstrate growing aptitude in higher education.

Demonstration of having taught before at university, ideally with experience in different teaching methods, is also valuable. However, anecdotal discussion of best educational practice through consideration of classroom experience is considered by most practitioners as a lower level skill compared to designing or enhancing taught content through principles of educational theory, which is associated with higher grades of the specialism.

Either completion of a PGCert in HE, or maintaining a portfolio of best practice written in relation to relevant educational theory, are the two routes to attaining recognition as a higher education professional by the UK Higher Education Academy (The National Committee of Inquiry into Higher Education, 1997), who award fellowship for such an achievement (FHEA). Higher levels of senior and principal fellowship (SFHEA/PFHEA) are then awarded for sustained work with impact at a cross-institutional or a national/international level.

20.8   Summary

The author has taught medical physics at all three university groupings and offers concluding thoughts as to the optimal strategies to succeed in a physics or engineering teaching career at each.

- *Russell Group universities* – These offer roles within a team of teaching professionals on a large programme, or a directing role on a smaller specialised programme. The latter opportunity can develop organisational and project management skills and maintain a strong professional identity as a medical physicist or clinical/biomedical engineer. However, job security is improved on larger courses, and working in a team can shelter the teacher-specialist from inevitable tensions between demands of researchers and educators at such institutes. Most Russell Group universities are currently worried about TEF and its outcomes.

- *Plateglass universities* – Teaching specialist staff will be expected to contribute heavily to programmes, with both large and small cohorts, as staff resource is more limited than at Russell Group universities. While kept busy, there is often good mentorship and career development opportunities as most educators at plateglass universities are suitably supported in their role because such HEIs market their university around offering a strong student experience. These universities are likely to do well in TEF.

- *Post-92 universities* – Teaching specialists often work among other professional experts to deliver foundation level courses. These serve as good developmental grounds for an early career and there are often approachable colleagues with teaching expertise to help gain familiarity with pedagogy. However, there are often reduced opportunities to work with advanced engineering or physics content. TEF performance is uncertain depending on the exact format of applied metrics (i.e. definition of employability, widening participation).

References

Department for Business, Innovation & Skills (2012) *The Browne Report. Securing a Sustainable Future for Higher Education.* Department for Business, Innovation & Skills, London. Available at https://www.gov.uk/government/uploads/system/uploads/attachment_data/file/422565/bis-10-1208-securing-sustainable-higher-education-browne-report.pdf

Department for Business, Innovation & Skills (2016) *Success as a Knowledge Economy: Teaching Excellence, Social Mobility and Student Choice.* Department for Business, Innovation & Skills, London. Available at https://www.gov.uk/government/uploads/system/uploads/attachment_data/file/523546/bis-16-265-success-as-a-knowledge-economy-web.pdf

Higher Education Academy (2016) *National Teaching Fellowship Scheme (NTFS).* Available at https://www.heacademy.ac.uk/recognition-accreditation/national-teaching-fellowship-scheme-ntfs

Higher Education Statistics Agency (2016) *Income and Expenditure of Higher Education Providers in 2014/15.* Available at https://www.hesa.ac.uk/news/03-03-2016/income-and-expenditure

Neves, J and Hillman, N. (2016) *Higher Education Academy – Higher Education Policy Institute (HEA-HEPI) Student Academic Experience Survey*. Available at http://www.hepi.ac.uk/wp-content/uploads/2016/06/Student-Academic-Experience-Survey-2016.pdf

Quality Assurance Agency (2008) *The Framework for Higher Education Qualifications in England, Wales and Northern Ireland*. August 2008. Available at http://www.qaa.ac.uk/en/Publications/Documents/Framework-Higher-Education-Qualifications-08.pdf

The National Committee of Inquiry into Higher Education (1997) *The Dearing Report. Higher Education in the Learning Society*. Her Majesty's Stationery Office, London.

Universities UK (2016) *The UK's Competitive Advantage*. Available at http://www.universitiesuk.ac.uk/International/Pages/uk-competitive-advantage.aspx

University of Birmingham (2015) *Shaping our Future: Birmingham 2015*. University of Birmingham. Available at http://www.birmingham.ac.uk/Documents/university/shaping-future.pdf

21 Teaching and learning in higher education

Alison Robinson-Canham, Sally Bradley and Catherine Lillie

21.1   Introduction

Historically it was assumed that those with a good knowledge of their subject and associated research credentials were fully equipped to teach in higher education. Such assumptions were challenged by the Dearing enquiry into Higher Education (HE) (Dearing, 1997) which recommended, among other things, the introduction of teaching qualifications for university teaching. In response, professional qualification and recognition mechanisms were established. The professional standards for teaching and supporting learning in higher education, the UK Professional Standards Framework (UKPSF) are now maintained by the Higher Education Academy (HEA) which awards Fellowships and accredits teaching qualifications delivered by higher education providers. At the time of writing, there are about 85 000 (January 2017) individuals recognised through HEA Fellowships and over 450 accredited training programmes and continuing professional development (CPD) schemes in the UK.

Public and government interest in the quality of HE teaching continues to increase and has never been more acute. Since the late 1990s, successive governments have promoted widening access to and participation in HE with 47% of 17- to 30-year-olds entering HE for the first time in 2013/14. This combined with the introduction of and increase in student tuition fees has led many families with no previous interest in HE to look closely at what they will get in return for the student debt incurred through university study. Regular satisfaction and engagement surveys (the National Student Survey (NSS) and UK Engagement Survey (UKES), for example) focus attention on student perceptions of their university experience. At the time of writing, the Higher Education White Paper heralds legislation to enact the 2015 Conservative election manifesto commitment to introduce a Teaching Excellence Framework (TEF). And so we have an environment in which more people than ever before are interested in the quality of teaching, the means by which that teaching quality might be measured, and the consequences for learners.

Read this chapter and reflect on what will make you the best educator you can be.

Find out more about...

- *The Dearing enquiry (https://www.leeds.ac.uk/educol/ncihe/).*

- *The UK Professional Standards Framework (UKPSF) (https://www.heacademy.ac.uk/ukpsf).*

- *UK HE participation rates (https://www.gov.uk/government/uploads/system/uploads/attachment_data/file/458034/HEIPR_PUBLICATION_2013-14.pdf).*

- *Policy Paper: Higher Education: Success as a Knowledge Economy - White Paper (https://www.gov.uk/government/publications/higher-education-success-as-a-knowledge-economy-white-paper).*

- *NSS and UKES (http://www.thestudentsurvey.com and https://www.heacademy.ac.uk/research/surveys/united-kingdom-engagement-survey-ukes).*

21.2 About the authors and this chapter

This chapter is written by three colleagues with extensive experience of working in higher education and professional bodies. We are coaches and educators who have developed standards and frameworks for professional and educational practice, and fostered and championed the means by which individuals can develop themselves and achieve formal recognition of their professional learning. Notwithstanding the intended audience for this book, we have been unashamedly 'not scientist' in our approach. Nor have we chosen to be traditionally didactic – there is no 'sage on the stage' here, dictating your learning journey. Our work is rooted in social theories of learning, which focus on how individuals learn and develop through living and being in their own social and professional contexts. We have also drawn on educational theories which explore the intellectual discomfort inherent in deep learning of complex concepts.

We have chosen to write in a conversational style to stimulate your active engagement with your own learning even as you read this chapter, and hope you will essentially engage in a quasi-conversation with yourself to explore issues around your own development and learning. There is a long and distinguished tradition of conversational dialogic learning, arguably traceable to Socratic discourse in ancient Greece, but here we want you to take personal responsibility for what and how you learn and so have coined the term 'monologic discourse' to capture this self-directed and reflective process. We hope you will embrace the potential of positive enquiry and celebrate your strengths, not just berate yourself with a deficit assumption that professional learning is a remedial route to simply filling knowledge and skills gaps.

Above all, your learning is unique to you, contextualised by your learning styles and preferences, the environment in which you work, your professional context, and subject, clinical or practice specialism. This chapter will help you better understand your own learning and development aspirations, understand the formal and informal professional learning available to you, and help you use your own 'monologic discourse' to navigate your learning journey. We have included 'Find out more about...', 'Stop and think!', and 'Practitioner perspective' boxes throughout the chapter to encourage you to make it real for you.

21.3　Why is teaching in HE so important? Values, drivers and equality of opportunity

As well as the increased demand for teaching qualifications, there are other drivers impacting on HE. Education has changed since the Dearing Enquiry – the rapid increase in participation triggered by government policy (the so-called 'massification' of HE), swings in vocational education, higher apprenticeships and the focus on employability, an increasing role for alternative and private providers, the growth of virtual learning technology and changing student expectations. With the impending introduction of the TEF, we are also increasingly concerned about how we measure 'good teaching'.

Some things, of course, do not change and for many, the role of university educator involves fostering the next generation of professionals, policymakers, intellectuals, innovators, scientists and creatives, thereby making a hugely significant contribution to our collective social, economic, cultural and intellectual future. For some, this sounds too big and overwhelming and they just want to get by and not make a mess of their 'teaching'. All these positions are entirely valid but it is worth taking a moment to think about your own motivation, drivers and values in relation to education.

Stop and think!

- *How do you know how good your teaching really is?*

- *How do you monitor the effectiveness of your education practice over time?*

- *What motivates your teaching practice?*

- *What is the ultimate purpose of your education practice?*

- *What do you believe is important about education?*

The UK Professional Standards Framework (UKPSF) captures the practice expectations for teaching and learning support in HE. As well as encouraging reflection of the knowledge and skills required to practice in HE, there are four professional values at the heart of the Framework.

UKPSF Values:

- *V1. Respect individual learners and diverse learning communities.*

- *V2. Promote participation in higher education and equality of opportunity for learners.*

- *V3. Use evidence-informed approaches and the outcomes from research, scholarship and continuing professional development.*

- *V4. Acknowledge the wider context in which higher education operates recognising the implications for professional practice.*

'Respect individual learners and diverse learning communities' makes explicit the need for all educators to consider our own actions and approaches in respect of students and their learning. Alongside this value sits the promotion of participation in higher education and 'equality of opportunity for learners'.

It is fair to say that the UKPSF values reflect some key changes in the social landscape since the pre-Dearing era – society has changed in that time, embracing more ambitious expectations of inclusivity, equality and diversity. These are issues which cannot be ignored especially within the scientific community where, for example, there is a known gender imbalance. The work of the Equality Challenge Unit (ECU, 2015) presents us with uncomfortable figures relating to gender, ethnicity and disability representation in HE.

According to the Department of Work and Pensions, disability prevalence is 16% amongst working age adults in the UK, with '14.9% of working age disabled people [holding] degree-level qualifications compared to 28.1% of working age non-disabled people'. The ECU data indicates that 46.6% of disabled staff work in Science, Engineering and Technology subjects (SET, more usually referred to as STEM subjects to include Mathematics), compared to 53.4% in non-SET/STEM subjects. Thus, across the academic domain, we should be asking whether we are creating barriers to disclosure and/or barriers to participation in SET/STEM academic careers, and whether this is a culture we experience, possibly even perpetuate, in our own education practice with students.

Discussions of inequality are often focused around gender, and the ECU report presents some disconcerting figures.

- The majority of all professors were men (77.6%). This was true across SET/STEM and non-SET/STEM subject areas and full- and part-time employment.

- This gender difference was most notable among full-time professors working in SET/STEM subject areas, where 81.8% were men.

- 34.8% of male academic staff earned over £50 000 compared with 20.1% of female academic staff (ECU, 2015).

In IPEM-related subject areas, the gender split is 24%:76% female/male in relation to gender by academic subject compared to 41%:58% across the wider portfolio of SET/STEM subjects, which would suggest that some consideration needs to be given to the equality culture within the medical science domain.

It may be useful to reflect on how this is changing or could change over time – the current gender imbalance amongst professors could derive from historic practices which have been challenged for some time, and yet there is a persistent narrative that the imbalance results from direct or indirect discrimination. Similarly, there is some evidence to suggest that, in NHS settings, the gender balance may be more equitable at early and mid-career stages, and so over time, this may translate into parity at senior levels. To achieve such equality, however, will require commitment from all concerned – in sectors where the gender balance and age profile of the workforce are more equal from the outset, there is still a marked prevalence of men in senior positions. For those charged with educating the next generation of scientists, or anyone else for that matter, the key issue to consider is how your own teaching practice fosters talent regardless of gender, ethnicity, or any other factor irrelevant to intellectual capability.

The Equality Act of 2010 identified nine 'protected characteristics' to describe some of the factors to consider when ensuring equality of opportunity. The Act also introduced the 'anticipatory duty' which required organisations and individuals within them to make necessary adjustments for the characteristics without relying on individuals to disclose the information themselves. Thus, there is an obligation for you as an educator to consider how your teaching practice is inclusive to all those protected by the legislation.

The nine legally protected characteristics are:

- age;
- disability;
- gender reassignment;
- marriage and civil partnership;
- pregnancy and maternity;
- race;
- religion and belief;

- sex;

- sexual orientation.

Resources and reasonable adjustments are required in both the face-to-face and online environments. Beyond the requirements of legislation, we should also consider the challenges faced by specific student groups, for example, first-generation students, parents with childcare responsibilities. It is no longer acceptable to say, '…but in my day.'

Stop and think!

- *Consider the diversity in your immediate team / working environment. Consider the roles, grades, disability (including hidden), gender and age profile. Why is this? What message does that convey to your students?*

- *Being completely honest with yourself, how inclusive is your practice?*

- *Do you unwittingly make assumptions about a student's potential on the basis of age, ethnicity, disability, gender, caring commitments or other factors unrelated to an individual's intellectual capacity to fulfil the professional requirements of your specialism?*

Find out more about … equality, diversity and inclusive practice from:

- Your institution's Educational Development/Academic Practice Unit.

- Your institution's equality and inclusion team.

- The Equality Challenge Unit (ECU) (http://www.ecu.ac.uk/).

- Access unconscious bias training, either from your HEI, or through another organisation.

- From IPEM (http://www.ipem.ac.uk/AboutIPEM/ EqualityDiversityandInclusion.aspx).

- **Find out more about …**
  - How your own background, personal experiences, societal stereotypes and cultural context can have an impact on your decisions and actions without you realising ('unconscious bias') at http://www.ecu.ac.uk/guidance-resources/employment-and-careers/staff-recruitment/unconscious-bias/

21.4   You as an educator

So, what sort of educator are you? You may want to think what your perspective of good teaching is – one size does not fit all. For example, teaching can be broken down into several elements: Transmission, Developmental, Apprenticeship, Nurturing, and Social Reform (Pratt, 2002). This will impact on how you teach. This chapter was originally conceptualised as 'teaching the teachers', so our starting assumption is that you are already engaged in supporting the learning of colleagues, students or patients, or are interested in doing so. You may be an experienced medical physicist, medical engineer, biomedical engineer, technologist or clinician who is moving into an academic role, or a researcher supporting more junior academic colleagues in their learning through doctoral supervision. You may be practicing in a clinical context where you are supporting trainees or working in inter-professional teams with different specialist colleagues, or conveying complex concepts and information to patients. All of these can be thought of as 'teaching' but we would encourage you to think of your role as that of an 'educator' – a more expansive term which avoids the traditional assumptions about the expert conveying predetermined knowledge and skills to passive recipients. There is undoubtedly an element of this inherent in education and training, especially where critical clinical judgements have to be made on the basis of sound scientific and technical knowledge, but we would argue that this is not the greatest challenge faced by educators expert in their subject matter. A far greater challenge resides in understanding the most effective learning processes for yourself and those whose learning you seek to support. Thus, our focus here is to support your development as an educator.

Understanding yourself as a learner is key to recognising habits, beliefs and behaviours which may be helpful to your role as an educator, either by fostering effective practices, or by helping you avoid unhelpful habits. Here are some questions and thoughts to get you started.

- How satisfied were you with how you were 'taught'? What would have made it better?

    - Anecdotal evidence has to be treated with caution, and learning and teaching practices are constantly evolving so that what you experienced as a learner is probably not the norm for current students. That said, everyone will have strong memories of their own education and notions of which teachers they would want to emulate, or not. Try to analyse what it was about your educational or professional heroes that made them such positive role models – and which you would hesitate to emulate!

- How did you learn the subject knowledge and skills in which you are now expert? What did you find difficult/easy?

- Often educators end up teaching subjects and skills in which they have become expert, and their journey to expertise has been aided by natural aptitude or fascination with the subject. As an educator, you cannot assume your learners will be equally adept or enamoured with the subject matter. Reflecting on how you learnt the things that you found hard or uninteresting may help you devise more engaging learning activities, and ultimately lead to more effective learning by your students.

- Were there things in your own education that seemed pointless, unsettling or anxiety-inducing at the time but subsequently have become part of your routine professional practice?

  - There is an educational theory of threshold concepts (Land *et al.*, 2005; Meyer and Land, 2003, 2005) which suggests there is some learning that is highly problematic and unsettling for learners as they grapple with complex concepts, the learning of which fundamentally alters their sense of their subject and their own relationship to it. Once you have crossed the threshold, and over time internalised the concept, it is hard to look back and understand from the perspective of new learners just how challenging that learning was. This can be a significant difficulty for experts seeking to support novices in their learning. Recent research (Bradley *et al.*, 2015) demonstrated that being passionate for your subject and engaging can be seen as more important than actual subject knowledge but being able to convey complex concepts in a way that is understandable is key to opening the door to the subject. Threshold concepts are key, taking students through that portal, metaphorically over the threshold, with something without which they cannot progress and which cannot be unlearnt once learnt. It is your challenge to put into words and practice what seems obvious to you as an experienced STEM practitioner but can be unsettling, confusing or dispiriting to the learner.

- Do you learn most readily through reading articles, doing practical experiments, looking at diagrams, or hearing someone speak about things?

  - There are theories of learning which suggest people have an inherent preference for learning in particular ways. VARK (Visual, Auditory, Read/write, Kinaesthetic) is one such model. Most people learn through a mixture but understanding where your strongest preference lies can help you overcome barriers to your own learning, and can help you nuance your teaching to accommodate the preferences demonstrated by your learners. You do not need to know the learning styles of all your individual learners but you can design learning materials and activities to accommodate the different styles you may encounter in your student group. Sometimes deliberately adopting a less comfortable mode of learning can reveal new insights or understanding that habitual learning behaviours mask.

- When you are learning something new, are you most attracted to understanding it from a theoretical perspective, thinking about it and puzzling it out for yourself, trying it out for real, or do you make the best of whatever situation presents itself?

  - Honey and Mumford's Learning Styles Questionnaire (Honey and Mumford, 1992, 2006) is widely used in higher education. It is more comprehensive than VARK and maps well against a learning cycle which encourages learners to consciously move from doing something, reflecting on it, trying something different, and re-evaluating, whilst recognising that different learners will be more or less comfortable at different points in the cycle. Recognising your own preferences and how those interact with other preferences can help you overcome frustration in working with learners and colleagues who have a different preference to your own – and help you avoid frustrating others!

- How confident do you feel about teaching?

  - How you feel about 'teaching' and 'education' is important. If you are enthusiastic and ideologically committed to sharing knowledge and skills, developing the next generation of academics and practitioners in your specialist field, generally supporting learners in fulfilling their intellectual and professional potential, you are more likely to cheerfully and positively engage with the learners you encounter. Even in this happy situation, you may find your learners frustrating sometimes. You may routinely feel quite negative about your teaching load – because it encroaches on time you would prefer to spend on research, or the balance of your workload may feel overwhelming and unmanageable, or you may not feel confident as an educator compared to how comfortable you feel as researcher or practitioner. Any ambivalence will show and your job will be harder if your anxieties or reluctance leak out. This is unfortunate for you and your learners so try to find some glimmer of interest either in the subject matter, the way you choose to deliver it, the impact you can have on your learners' success, or simply in the need to do a job as well as possible. The important thing is to be honest with yourself, not berate yourself for your authentic feelings, and to develop personal strategies for drawing on your subject, practice or research enthusiasms so you can inspire that enthusiasm in others. You could try using your own recent publications as teaching aids or anonymised examples of real patients in clinical settings; remember why you entered your chosen profession, or pursued your specialism, or how you handled challenges in your own learning; find ways of involving your students in your current research, or discuss with your students your and their motivation for the subject. Essentially it comes down to finding some common ground so you can foster reciprocal empathy between student and teacher so both can learn from each other.

- How does education/teaching/supporting learning relate to other aspects of your work?

  - Research-informed teaching is a key tenet of higher education, the use of an evidenced approach, how you use your personal scholarship, evidence from research, from your personal engagement in professional learning and from your own professional practice. This may be implicit in your teaching practice but have you considered how your research or professional practice influences the content of your teaching? Equally, you may want to consider how your teaching influences and shapes some of your research. Are you sharing this practice with your colleagues? Also, to what extent are you fostering research and professional practice skills in your students?

Stop and think!

- *Why did you enter this field? What inspired you to your practice? Was it a person or a concept? Who was your most inspirational educator? What did they do to enthuse you? What was the environment like?*

- *Think about something which you learnt as a student, which is key to your practice. This may be something which you use on a day-to-day basis without even thinking.*

- *How could you explain this to someone?*

- *This is what your inspirational teachers will have achieved, perhaps unknowingly.*

Find out more about...

- Learning styles: VARK (http://vark-learn.com/the-vark-questionnaire/).

- Learning styles: Honey and Mumford from your staff and education development unit.

- Communities of Practice and being novice and expert (http://www.learning-theories.com/communities-of-practice-lave-and-wenger.html).

- Threshold concepts (http://www.ee.ucl.ac.uk/~mflanaga/thresholds.html).

Practitioner perspective

Why did you go into HE teaching?

I had an inspirational teacher when training as a radiographer. And I wanted to be like him!

At that time, it wasn't a degree but was very clinically focused. In the 2nd year, a young student teacher joined the programme. He just inspired me, his energy and enthusiasm for the subject. From then on. I knew that radiography was the career for me, but I wanted to go into teaching with the ultimate goal to go into education.

But I knew I needed a grounding in clinical practice, so worked toward being a clinical educator. I wanted to be a good educator.

What happened to the junior teacher?

He became Head of the Radiography programme at Bradford University. I later ended up working for him as a clinical teacher and more recently worked alongside him in a national education forum (which I now Chair!).

What led you to Radiography?

At 17/18, I did not really know what I wanted to do. Then I had an accident resulting in a broken arm and head injury. Whilst in hospital, I looked up at the radiographer doing the skull examination and decided there and then that I was going to be a radiographer. I liked science, patient care and new technologies. Career chosen!

Professor Julie Nightingale, University of Salford

21.5   Wearing multiple hats and managing your own discomfort

As an educator, you wear a number of different hats on any given day.

- sometimes the expert; other times the novice;

- sometimes communicating with peers in technical language you all understand; other times explaining complex terms to vulnerable or confused patients with no prior understanding of the area;

- sometimes imparting knowledge to students who are keen to engage; other times feeling that your pearls of wisdom are falling on deaf ears;

- sometimes acting as coach, mentor or role model to less experienced colleagues; other times needing this type of advice and guidance from others;

- sometimes lecturing to 200 students; other times leading 1-1 postgraduate supervision meetings.

Stop and think!

Consider the different hats you have worn in the last week. Which did you enjoy the most? Why? Which did you find easiest? Why?

Practitioner perspective

'None of us will reach a stage where we have learnt everything there is to know about the job and we all must learn new skills and knowledge, and also act as educators to those around us at each stage of our career. In my case, my latter NHS years were a combination of NHS service work in radiation protection and diagnostic radiology, and research work. The research was performed within the NHS and was mostly my own hands-on work. The two types of work were very different; risk averse, not that creative, often repetitive work on radiation protection and QA, with a culture of not wandering outside of the box. On the other hand, research was creative, sometimes haphazard in the thought processes one had to engage, and the rewards were greatest when one deliberately wandered outside of the box into new unexplored areas. I struggled with holding these two hats at the same time. Fortunately, these activities took place in different hospitals and I biked between the two sites, morphing from one role to another between the two, to the point where I was unable to deal with queries in one area if these were presented in the wrong hospital.'

Professor Peter Hoskins, University of Edinburgh

You will need to be highly adaptable to meet all these different expectations, and won't always get it right. But a good starting point for any communication is the audience – what do they know already, what do they need to know and what is the best way of telling them? Strengthening your strengths is also a much more effective way of developing yourself than focusing on weaknesses, so identifying and using your strengths in all of the different areas of your professional practice is a quick win. Creators of the model of situational leadership posited that the best leaders will adapt their style to those they are seeking to lead (Blanchard, 1987; Hersey *et al.*, 1979). Having this in mind may help you to decide how to act and how to get the best out of others in any given situation.

21.6   Methods of teaching and learning

As we have already acknowledged, the world in which higher education sits has probably changed since you 'read' (deliberate wording) for your undergraduate degree or undertook your professional training. Government agendas have influenced the way we teach and students learn. Employer involvement and expectations, reduction in part-time study, increased focus on employability and internationalisation of the curriculum, and the demographic of both undergraduate and postgraduate courses have all evolved over the last 15–20 years.

Perhaps the most significant change has been in the expectations and 'consumer' behaviour of students who now not only invest their intellectual and scholarly energy in their education, but also incur significant debt in pursuit of the much promoted, and for some almost mythical, 'graduate earnings premium'. So, how much teaching do students expect to receive? The introduction of student fees led to a sharp focus on contact hours, the QAA provides a list of methods for learning and teaching (QAA, 2011, p. 6):

- lecture;
- seminar;
- tutorial;
- project supervision;
- demonstration;
- practical classes and workshops;
- supervised time in studio/workshop;
- time in which students work independently but under supervision;
- fieldwork;
- practical work conducted at an external site;
- external visit;
- work-based learning.

Whilst the list refers to workshops, we would suggest that this also includes laboratory work. Each method requires different skills, from formal lecture to facilitating work-based learning. This language and codification of 'contact hours' can appear to diminish the role and responsibility of the individual as an autonomous learner, so you may want to consider how your education practice equips your learners for the independent and self-directed study which arguably prepares them for their employment after graduation.

It is no longer the case that an expert in their subject is also assumed to be an excellent teacher; teaching is not about filling empty vessels with knowledge but about developing skills to analyse and evaluate and to apply knowledge in different contexts. Which raises the question, what skills do staff need to engage students within the range of environments listed above as well as new and emerging environments such as social media?

Find out more about...
- Contact hours from QAA (http://www.qaa.ac.uk/en/Publications/Documents/contact-hours.pdf)

Earlier in this chapter, we discussed understanding your own learning preferences and being sensitive to the learning preferences of others. Such an approach usually advocates interactivity and the use of pictures and activities rather than just text. Most important is that you are alert to the learning needs of your students, acknowledging and respecting their preferences, being alert to and honest about your own habits and educational behaviours, and ultimately flexing your practice to your specific circumstances. Sometimes less is more and you just need to be pragmatic!

Practitioner perspective

'Students in higher education are all too often presented with slides cluttered with information by academics attempting to move away from dense text. The major issue being that the layout of the information presented is far too chaotic – with unnecessary visual matter, such as photos, cartoons and other such graphics which sometimes distract from the message of the slide and could appear overwhelming to some students.

My concern when designing or adapting course material is for students who have additional learning needs; international students make up a large percentage of postgraduates studying in the faculty and are unaccustomed to information presented in this way, where they have to fight through the distractions to understand which is the main message being conveyed. In addition, for students with English as an additional language, large sections of text are not always appropriate in time-restricted teaching, as it would naturally take longer to progress through the information. This would result in a progressive loss of attention and ultimately remove the benefit of the programme. As such, I design course material to effectively state information in a more direct manner to ensure accessibility to all material provided. This means clarifying jargon and removing UK cultural references where these are not appropriate.'

Dr Sarah Mohammad-Qureshi, University of Manchester

Mohammad-Qureshi's perspective above illustrates how notions of 'good practice' can be contested and shift with time and fashion, varying between different types of learning environment, and cultural contexts. Even if you do not agree with everything that is promoted as good practice, reflecting on how different ideas may or may not apply to your own situation will open up new avenues for your consideration – as Henry Ford said 'If you always do what you've always done, you'll always get what you've always got.'

So, within the traditional learning environments – lecture theatre, tutorial or seminar – what skills do you need? We've captured just a few of the perennial dilemmas and tropes of 'teaching' and invite you to reflect on when and how they are significant for your practice.

For example, much is said about **presentation and communications skills**, some of which is gently contested in the practitioner perspective provided by Dr Mohammad-Qureshi. Start by thinking about your own communication habits – skilled and engaging communication or death by PowerPoint? Do you cram your book chapter on a set of slides and read the content? Do you provide bullet points, which are really just a prompt for you rather than a set of slides which could be used as a revision tool?

Do you provide a transcript or notes in addition to the slides so that the students have something to read outside of the lecture? As already addressed, your students will have preferred learning styles, but that is not to say they cannot learn through other styles. How do you ensure that all of the styles are covered, or do you concentrate on your own preferred way of learning, 'well, it worked for me' style of teaching?

You may have thrived academically despite unimaginative teaching, or because of inspiring educators, but even so, practice and the world changes. Realistically, when was the last time you sat and listened to something, uninterrupted for 50 minutes (without reaching for your phone to email, text or tweet)? And yet many of us routinely resort to this familiar model (Atherton, 2013). Does it feel safer to escape the lecture or tutorial leaving no time for questions? Are lectures a waste of time and simply a way of getting information from the lecturer's notes to the students' notes without going through the brains of either, with the learning actually done on one's own when one has time to think? Why not try breaking the session up? Asking questions, or a show of hands as a poll, are easy ways to keep students engaged. Encourage students to reflect on their own and consolidate their own learning If you are brave, you could attempt some group work, yes, even in a lecture theatre!

Oh, and then of course there is that thing called '**peer observation**'; done well this can be one of the most developmental experiences you have. What can you learn from observing someone else teach? If you get the chance, have a look at how a historian or creative writer teaches, you may be surprised at what you can pick up.

Project supervision, especially at undergraduate level, can be seen as either tedious or incredibly exciting. Every undergraduate is potentially your next research associate, so think about what skills you want them to have. What skills do you need to develop your own 'assistant'? Students as researchers is not a new concept; Angela Brew has been promoting it for many years. More recently, the Higher Education Academy has commissioned work such as 'Students as researchers: Supporting undergraduate research in the disciplines in higher education' (Brew *et al.*, 2011).

Chickering and Gamson (1989) came up with seven principles of good practice in HE teaching and learning. Good practice:

- encourages student–faculty contact;
- encourages cooperation among students;
- encourages active learning;
- gives prompt feedback;
- emphasises time on task;
- communicates high expectations;
- respects diverse talents and ways of learning.

Let's unpick some of these and see how relevant they are today.

Contact with students is vital, so how accessible are you outside the lab, lecture theatre, seminar room? Social media can be the bane of our lives but emails, virtual learning environments (VLE) and social media probably make you more accessible than your lecturers when you were at university. But how effective are they for 'teaching'? How can you harness social media to keep students engaged? When sharing something you have read or seen in the news, it's easy to share a link through Twitter or Facebook or news area within your VLE. You do not have to be physically present to keep in touch.

Many IPEM-related educators will be working with trainees in work-place learning situations, with a host of associated preoccupations and distractions. For your students, the clinical or work-place context is an invaluable learning opportunity; for your colleagues or patients, those students could be a source of concern, anxiety or annoyance. It will be helpful for you to consider how you might reconcile your education practice with your 'day job'. What will it be safe and reasonable to allow students to do? How can you protect time for supporting your work-place learners? Being very clear about the learning outcomes students are expected to meet in the work-place will be essential to managing these situations. You never know, your students might actually be useful and may have been introduced to more up-to-date theories and ideas than you are using day-to-day in a clinical setting. You can provide them with a reality check and they can keep you on your toes!

Find out more about…. engaging learning and teaching through social media,

- Have a look at 'LTHE Twitter Chat' or the HEA Twitter feeds or blogs.

The pedagogic potential of social media and new technologies is undeniable. You will, however, need to establish appropriate boundaries to protect your own time, work and life balance, and privacy, whilst still being accessible to your students.

Cooperation – students learning together, especially through problem-based learning provides opportunities to explore and discover whilst developing key employability skills. Suitably scaffolded and supported student projects may be the perfect vehicle for safely trialling the professional problem solving skills essential to the domain of medical science and engineering. James Atherton provides some key pointers with regard to learning in groups (Atherton, 2013) and Jaques and Salmon (2007) cover face-to-face groups and virtual groups.

Active learning is a term which epitomises many of the social theories of learning with emphasis on learning through being and doing in real situations. So how do you get your students to engage actively in their learning? A number of educational

213

theorists and commentators write on this topic, and the CBI recently articulated graduate employability to include: 'self-management – readiness to accept responsibility, flexibility, resilience, self-starting, appropriate assertiveness, time management, readiness to improve own performance based on feedback/reflective learning...a positive attitude: a 'can-do' approach, a readiness to take part and contribute, openness to new ideas and a drive to make these happen' (CBI/UUK, 2009).

Barnett points out that there has been 'a strong emphasis on developing flexible teaching and provision without a similar focus on the development of students as flexible learners' (Barnett, 2014). Ryan and Tilbury (2014) take this further and suggest that flexible learning ought to entail developing student capabilities to operate flexibly: 'to think, act, live and work differently in complex, uncertain and changeable scenarios'. Their argument is that this might push us to consider flexible pedagogies which promote the development of flexible capability in students. Why not have a look at some alternative ways of engaging students such as 'Using game-based learning to engage students with physics: how successful could 'Junkyard Physics' be?' (https://www.heacademy.ac.uk/resources/detail/stem-conference-2014/Post_event_resources/Physical_sciences/Using_game-based#sthash.ygumGWR2.dpuf).

A step too far? Come back to what cognitive skills you want your future research associate to have, would this help develop those skills?

Feedback – Anyone in higher education will have heard the discussions with regard to the National Student Survey (NSS) and the impact of the scores for feedback. It is worth thinking a bit about why students might feel negative about feedback.

First, feedback isn't always obvious. Students do not always realise how much of their interaction with teachers is actually feedback. Feedback does not have to be a formally written response to a formally assessed task. The routine dialogue between teacher and learner can be a rich medium for exploring ideas and sense-checking or correcting understanding. The onus is on the teacher to exploit these opportunities for providing informal feedback.

Second, assessment can be either **formative** (a task set to promote learning and check progress towards learning outcomes, which identifies areas for further work, the assessment of which does not usually contribute significantly or at all to marks or results), or **summative** (a task set to formally assess whether learning outcomes have been met and which does contribute to marks or results). Sometimes an activity which does not contribute to marks or results is not treated as seriously by either teacher or learner, and so the feedback arising from formative assessment – the feedback most useful to improving learning – is not as useful or as valued as it could be. As the teacher, remember that formative assessment and feedback are critical components of effective teaching, and offer rich opportunities for influencing improved learning and ultimately results.

214

Third, there is an emotional response to receiving feedback, whether good or bad. Your students have invested time and energy into producing a piece of work or fulfilling a task, and will be keen to know what you, as the expert, think about their work and how they can improve their performance. If they haven't invested the time you would like, they will need to hear that as well. Sometimes the feedback isn't comfortable and you may need to allow time for it to sink in before you can have a really useful conversation with the student.

So, how do you give effective feedback? It isn't just that written comment on a script or lab book.

Critically, the feedback needs to be perceived as useful in order for students to engage with it. Formative feedback should provide the evidence of where they are now and how to bridge the gap to the goal. Summative feedback needs to be meaningful and actionable, for instance, if feedback is given at the end of a module, how can that feed forward into the next modules? It needs to be timely; no point in giving feedback which will help in exams if it is received on the day before the exam!

Stop and think!

…about the different ways you give feedback

Question and answer, model answers, audio feedback, multiple choice questions, peer marking, drafts, conversation in the corridor… Students do not always recognise this as feedback, so think how you can get students to recognise that it isn't just written comments.

Feedback matters – think about the emotional fallout of feedback, how the recipient feels. Say what needs to be said in the most constructive way possible. Can't remember getting negative feedback at university? Maybe not, but if you have ever had a research paper rejected, you will know what we mean.

Find out more about…

- Feedback: For more ideas on how to actively engage students in feedback: 10 Strategies to Engage Students with Feedback (https://www.heacademy.ac.uk/sites/default/files/resources/10_strategies_to_engage_students_with_feedback.pdf).

- NSS and UKES (http://www.thestudentsurvey.com and https://www.heacademy.ac.uk/institutions/surveys/uk-engagement-survey).

21.7 Developing yourself as an educator

We have just scratched the surface of some of the skills and techniques needed for effective teaching. Hopefully you will have started to question some of your own assumptions about how you and others learn, and to think about the skills you have already and those you might want to develop. Most UK HE institutions now offer a Postgraduate Certificate in learning and teaching or academic practice, or equivalent. You may have been offered, or even required to undertake, such a programme and felt ambivalent about its value to you. We would urge you to grasp the opportunity with both hands and with an open mind. Yes, it will take up time, and yes, it may on occasion feel like a complete distraction from other aspects of your work, but it will also give you time and space dedicated to in-depth consideration of the issues we have barely introduced here. If you have been in education practice, teaching for some time, you may wish to seek professional recognition by gaining an HEA Fellowship – you will have opportunity to reflect on what makes your teaching and learner support effective, gain recognition for the effort you invest in such activity, and gain access to a community of about 85 000 (January 2017) individual practitioners with whom you can debate and consider current thinking in pedagogic practice. Either option will challenge you to think about your practice, why you teach in a certain way, how you know it works and what underpins your practice. 'We've always done it like that' is no longer acceptable.

Practitioner perspective

In 2010, I was asked to put together a new MSc in Biomechanics. I welcomed the new challenge and was acutely aware that I had had no previous university teaching experience, arising from the fact that for most of my career, I was in the NHS. Fortunately, there was a PGCert in University Teaching which I could undertake so, in parallel, I was both teacher and leader on my MSc, and student on my PGCert. I gained new insights and understandings into the way people learn, but also into working across the whole University as fellow students were from the whole range of areas within the University. I found the PGCert to be a superb experience, undoubtedly one of the best experiences I have had in my time at Edinburgh University.

Professor Peter Hoskins, University of Edinburgh

Find out more about…

- Postgraduate certificates in teaching and learning or academic practice from your university's HR or educational development unit.

- The Higher Education Academy, professional recognition for your teaching practice, and ad hoc training and development opportunities, including 'New to Teaching in STEM' (www.heacademy.ac.uk).

Your development as an educator will be career-long. If you consider how much the sector has changed since the Millennium in ways it would have been impossible to predict previously, just imagine how much things will continue to change in the next few years. These changes will be in:

- the subject matter;
- the context in which you teach;
- the expectations and prior knowledge of those you teach, across the board from undergraduate to postgraduate research students;
- how you are measured on your teaching quality;
- the way in which you conduct your teaching;
- modes of delivery and assessment.

This does not just relate to online learning or new technologies either, but also to the development of different approaches and theories, and the different expectations HEIs set with regard to teaching and learning. At the very least, your professional learning will need to keep pace with these changes. How are you going to achieve this? It sounds a lot to weigh up, but the good news is that, as a professional innovator and problem-solver with an analytical mind, you are already well-placed to meet some of these challenges.

We are all of us on a career pathway or professional journey. Whilst once these were linear and clearly defined, careers today are much more likely to 'career' about and be unique for each individual. This gives you great opportunities to shape what you do and how you do it. The professional choices we have made and roles we have taken on may not follow a pattern, but, looking back, we can often see the threads of our interests, passions and strengths woven through all our roles.

> **Stop and think!**
>
> *What sort of educator will you be and what next steps might you take?*
>
> *Consider investing some time in a proper career review, but meanwhile ask yourself:*
>
> *Where have I come from?*
>
> *Where am I now?*
>
> *Where am I going?*
>
> *Have I got a plan?*
>
> *Is it my plan or someone else's?*

None of us know exactly what the future looks like. Teaching and education may be just one aspect of your current portfolio, and may increase or diminish in importance for you over time. Whenever you are working with learners, however, whether in labs, clinics, or classrooms, you do have responsibility for the quality of the learning experience you provide. This may sound daunting but, as we've started to explore in this chapter, you can be prepared by knowing your drivers, your personal capacity and capabilities, and by building a professional portfolio that plays to your interests. This may not be in a way that you had expected, or following the same path as your colleagues, but being open to new possibilities can take your career in exciting new directions and help meet some of the new challenges which are inevitably on their way.

Enjoy the journey!

References

Atherton, J. (2013) Learning and Teaching; Lectures. Retrieved 12 May 2016, from http://www.learningandteaching.info/teaching/lecture.htm

Barnett, R. (2014) *Conditions of Flexibility: Securing a More Responsive Higher Education System.* HEA, York.

Blanchard, K.H. (1987) *Leadership and the One Minute Manager.* Fontana.

Bradley, S., Kirby, E. and Madriaga, M. (2015) What students value as inspirational and transformative teaching. *Innovations in Education and Teaching International,* Vol. 52, Issue 3, 231–242.

Brew, A., Boud, D. and Un Namgung, S. (2011) Influences on the formation of academics: the role of the doctorate and structured development opportunities. *Studies in Continuing Education,* Vol. 33, Issue 1, 51–66.

CBI/UUK (2009) *Future Fit: Preparing Graduates for the World of Work*. CBI, London.

Chickering, A.W. and Gamson, Z.F. (1989) Seven principles for good practice in undergraduate education. *Biochemical Education*, Vol. 17, Issue 3, 140–141.

Dearing, R. (1997) Higher Education in the Learning Society, the Final Report of the National Committee of Inquiry into Higher Education. Available from http://www.leeds.ac.uk/educol/ncihe/ [accessed on 05/11/2014].

ECU (2015) Equality in Higher Education: Statistical Report 2015. ECU, London.

Hersey, P., Blanchard, K. and Natemeyer, W. (1979) Situational leadership, perception, and the impact of power. *Group & Organization Studies (pre-1986)*, Vol. 4, Issue 4, 418.

Honey, P. and Mumford, A. (1992) *The Manual of Learning Styles*. Peter Honey Publications Ltd, Maidenhead.

Honey, P. and Mumford, A. (2006) *The Learning Styles Questionnaire: 80-Item Version*. Peter Honey Publications Ltd, Maidenhead.

Jaques, D. and Salmon, G. (2007) *Learning in Groups: A Handbook for Face-to-Face and Online Environments*. Routledge, London.

Land, R., Cousin, G., Meyer, J.H.F. and Davies, P. (2005) Threshold concepts and troublesome knowledge (3): implications for course design and evaluation. In: *Improving Student Learning: Diversity and Inclusivity*, edited by C. Rust. Oxford Centre for Staff and Learning Development, Oxford.

Meyer, J. and Land, R. (2003) Threshold concepts and troublesome knowledge (1): linkages to ways of thinking and practising. In: *Improving Student Learning: 10 Years On*, edited by C. Rust. Oxford Centre for Staff and Learning Development, Oxford.

Meyer, J.H.F. and Land, R. (2005) Threshold concepts and troublesome knowledge (2): Epistemological considerations and a conceptual framework for teaching and learning. *Higher Education: The International Journal of Higher Education and Educational Planning*, Vol. 49, Issue 3, 373–388.

Pratt, D.D. (2002) Good Teaching: One Size Fits All? [Article]. *New Directions for Adult & Continuing Education*, Spring 2002, no. 93, 5.

QAA (2011) Explaining Contact Hours. *Guidance for Institutions Providing Public Information about Higher Education in the UK*. The Quality Assurance Agency for Higher Education. Available at http://www.qaa.ac.uk/en/Publications/Documents/contact-hours.pdf

Ryan, A. and Tilbury, D. (2014) *Flexible Pedagogies: New Pedagogical Ideas*. HEA, York.

Lightning Source UK Ltd.
Milton Keynes UK
UKOW07f2156210917
309650UK00003B/28/P